Concise Guide to

ENVIRONMENTAL DEFINITIONS, CONVERSIONS, and FORMULAE

Edward W. Finucane, PE, QEP, CSP, CIH

CRC Press
Taylor & Francis Group
Boca Raton London New York

CRC Press is an imprint of the
Taylor & Francis Group, an **informa** business

Acquiring Editor: Ken McCombs
Project Editor: Albert W. Starkweather, Jr.
Cover design: Dawn Boyd

CRC Press
Taylor & Francis Group
6000 Broken Sound Parkway NW, Suite 300
Boca Raton, FL 33487-2742

First issued in hardback 2017

© 1999 by Taylor & Francis Group, LLC
CRC Press is an imprint of Taylor & Francis Group, an Informa business

No claim to original U.S. Government works

ISBN-13: 978-1-56670-315-4 (pbk)
ISBN-13: 978-1-138-43489-9 (hbk)

Library of Congress Cataloging-in-Publication Data

Finucane, Edward W.
   Concise guide to environmental definitions, conversions, and formulae /
   Edward W. Finucane
        p.   cm.
     Includes index.
     ISBN 1-56670-315-8 (alk. paper)
        1. Enivronmental engineering—Handbooks, manuals, etc. 2. Industrial hygiene—Handbooks, manuals, etc. 3. Industrial safety—Handbooks, manuals, etc. 4. Environmental engineering—Problems, exercises, etc. 5. Industrial safety—Handbooks, manuals, etc. 6. Industrial safety—Problems, exercises, etc. I. Title.
     TD145.F55 1998
   628—dc2   98-30925
                                                                              CIP

Library of Congress Card Number 98-30925

Visit the Taylor & Francis Web site at
http://www.taylorandfrancis.com

and the CRC Press Web site at
http://www.crcpress.com

Dedication

With Pride and Gratitude,
this work is dedicated to my family:
*To my Wife, **Gladys**,*
who deserves to be identified
as the book's co-author;
*and to my two Sons, **Phillip** and **Ryan**,*
who, every day, make me proud to
be their Father.

Author

Edward W. "Ted" Finucane was born in San Francisco and raised in Stockton, CA. He has earned degrees in Engineering from Stanford University and in Business from Golden Gate University. Professionally, Mr. Finucane has been involved in both the Environmental and the Occupational Safety and Health fields for more than 30 years. During the last 18 years, he has operated his own professional consulting company, High Tech Enterprises, out of offices in Stockton. He is a Registered Professional Engineer (PE), a Qualified Environmental Professional (QEP), a Certified Safety Professional (CSP), and a Certified Industrial Hygienist — Comprehensive Practice (CIH).

He has had extensive experience in the areas of: ambient gas analysis, gas analyzer calibration, indoor air quality, ventilation, noise and sound, heat and cold stress, health physics, and in the general area of hazardous wastes.

For the past several years, he has served on the faculty of the twice yearly course, *Comprehensive Review of Industrial Hygiene*, offered jointly by the Center for Occupational and Environmental Health (University of California at Berkeley, California) and the Northern California Section of the American Industrial Hygiene Association.

Mr. Finucane's e-mail address is: **ted@hi-tech-ent.com**.

High Tech Enterprises' home page is: **http://www.hi-tech-ent.com**.

Preface

This book is intended to serve these purposes:

1. To function as a ready reference for the Occupational Safety and Health Professional, the Industrial Hygienist, and/or the Environmental Engineer. Such an individual, in the normal development of his or her career, will likely have specialized in some relatively specific subarea of one of these overall disciplines. For such an individual, there will likely be occasions when a professional or job related problem or situation will arise, one that falls within the general domain of Occupational Safety and Health, Industrial Hygiene, or the Environment, but, at the same time, is outside of his/her area of principal focus and competence. As a result of this, such situations will, therefore, be not immediately familiar to him or her. For such cases, this Reference Source, hopefully, will provide a simple path toward the answer.

2. To function as a useful Reference Source, Study Guide, or refresher to anyone who is preparing to take either the Core or the Comprehensive Examination for Certification as an Industrial Hygienist, a Safety Professional, an Environmental Engineer, or an Environmental Professional.

3. Finally, to assist Students who have embarked on a course of study in one of these disciplines. As a fairly concise compilation of most the important mathematical relationships and definitions that these Students will be called upon to utilize as they progress in their profession, it is hoped that this group, too, may find this work to be of some value.

Should any reader wish to pass on comments or suggestions as to any aspect of the contents of this volume, I can be reached at any of the following locations and/or listings:

High Tech Enterprises

P. O. Box 7835

Stockton, CA 95267-0835

Telephone (209) 473-1113

Fax (209) 473-1114

E-mail ted@hi-tech-ent.com

Home Page http://www.hi-tech-ent.com

Finally, I would like to compliment and thank any reader who has taken the trouble to wade through all the foregoing commentary. I hope it will be helpful as you progress in your studies or your career. Good luck as you put this volume into practical use.

Edward W. Finucane
PE, QEP, CSP, CIH

Acknowledgments

In authoring this text, I would like to thank a number of people without whose insights and guidance this work still would not be complete. Included in this group are professional associates, colleagues, friends, and even family members — in each case, individuals whose perspectives and opinions were very important to me.

First, I would like to thank Kenneth P. McCombs, Acquisitions Editor, and his associate, Susan Alfieri, Production Manager, both of whom work with my publisher, Lewis Publishers/CRC Press, in New York City. Ken, in particular, was the driving force behind pushing this work to its completion. In addition, Project Editors Mimi Williams, who worked on the book from which this volume is drawn — *Definitions, Conversions, and Calculations for Occupational Safety and Health Professionals*, Second Editon, published earlier this year, and Albert Starkweather, who worked on this volume — both also with Lewis Publishers/CRC Press in their Boca Raton, FL, offices — both made very important and substantial contributions to this work.

For sharing his significant experience in gas analyzer calibrations and standards, I would like to thank Wayne A. Jalenak, Ph.D., of Maynard, MA. His comments and insights on the material in the chapter covering Standards and Calibrations were invaluable.

For their contributions on the chapter covering Ionizing and Non-Ionizing Radiation, I would like to thank my brother, James S. Finucane, Ph.D., of Bethesda, MD, and David Baron, PE, of Minneapolis, MN. Jim's help with the overall structure of this section and Dave's unique contributions in the area of non-ionizing radiation were, in each case, absolutely vital to the development of the information in this chapter.

For sharing his expertise in the area of Statistics and Probability, I would like to acknowledge the contributions of William R. Hill of Albuquerque, NM. Bill's comments and suggestions were vital in clarifying the various difficult relationships that are discussed in this chapter.

David L. Williams of Santa Clara, CA, and Joel E. Johnson of Wilmington, DE, each provided their very valuable perspectives on the content of Appendix C, the section that covers the Atmosphere.

For the knowledge and inspiration he provided me, I would like to acknowledge my teacher, Professor Andrew J. Galambos, whose work in the physical and volitional sciences has provided me with the principal foundation upon which my own professional life has been based.

Last, but most certainly not least, I would like to acknowledge and thank my wife, Gladys. In spite of the fact that her formal education included neither the environment nor the area of occupational safety and health, she proofread the entire text, and in doing so was able to identify numerous areas where my descriptions required clarification, where I had omitted important data, etc. Needless to say, to the extent that the material in this book is understandable to its readers, much of the credit must go to her.

Table of Contents

Chapter 1   **The Basic Parameters and Laws
of Physics and Chemistry**

Chapter 2   **Standards And Calibrations**

Chapter 4   Ventilation

Chapter 1
The Basic Parameters and Laws of Physics and Chemistry

The basic parameters or measurements of the physical sciences (including all of the most common and widely used units that apply to each) will be identified and described in this chapter. In addition, the fundamental laws that find significant usage in the overall study and practice of industrial hygiene/occupational safety and health also will be covered in detail.

Relevant Definitions
Basic Units

There are seven basic or fundamental units of measure in current use today throughout the world. The most widely recognized set of these units, known as the **International System of Units** (*SI*), was initially adopted in 1960, and is reviewed and amended, as deemed necessary, at one of the General Conferences on Weights and Measures, an international meeting that convenes periodically. In addition, there are two common "metric" systems, referred to in the text that follows as the **MKS System** (meters, kilograms, and seconds), and the **CGS System** (centimeters, grams, and seconds), as well as the "nonmetric" **English System** (obsolete almost everywhere on Earth, except in the United States). Each of these Systems of Units will be covered.

Length

Length is the extent or distance from one end of an object to the other, or a distance in space from one clearly identified point to any other such point.

In the International System of Units, the basic unit of *length* is the **meter**, which has been defined as the length of path traveled by light in a vacuum during a time interval of 1/299,792,458 of a second.

In the MKS System, the basic unit of length is the **meter**. In the CGS System, the basic unit of length is the **centimeter**. In the English System, the basic units of length can be either the **foot**, the **inch**, or the **yard**.

Mass

In physics, *mass* is the measure of a body's resistance to acceleration. The *mass* of any object is different than, but proportional to, its weight — weight is a force of attraction that exists between the object being considered and any other proximate massive object (i.e., the earth).

In the International System of Units, the basic unit of *mass* is the **kilogram**, which has been defined as being equal to the mass of the international prototype of the kilogram.

In the MKS System, the basic unit of mass is the **kilogram**. In the CGS System, the basic unit of mass is the **gram**. In the English System, the basic unit of mass is the **slug**. In addition, although its definition as a *mass* is very confusing (largely because this same unit also serves as the basic unit of *force* for this system), a commonly used unit of *mass* is the **pound mass**, which has been defined to be that mass having an *exact weight* of 1.0 pound force when measured on the earth at sea level (i.e., a one pound mass would exert a downward force of one pound force on a scale situated at sea level).

Time

Time is the interval that occurs or exists between any two clearly identified events. In contrast to the situation with respect to length and mass, the basic unit of time is the same for all Systems of Units.

Until recently, the basic unit of *time* was defined to be the length of a *mean solar day*. Now, however, the basic unit of *time* is the **second**, which had been previously defined to be 1/86,400 of one *mean solar day*, but is now more precisely defined and quantified, according to the International System of Units, the MKS System, the CGS System, and the English System as the duration of 9,192,631,770 periods of the radiation corresponding to the transition between the two hyperfine levels of the ground state of the $^{133}_{55}$Cs atom.

Temperature

Likewise, *temperature* was not considered to be a basic unit of measure until more recent times. It has now become regarded as one of the fundamental seven units of measure. Simply, *temperature* is a measure of the relative "hotness" or "coldness" of any object or system.

In the International System of Units, the basic unit of *temperature* is the **degree Kelvin**, which — as the unit of thermodynamic *temperature* — is a fraction (specifically, 1/273.16) of the thermodynamic temperature of the triple point of water.

Quantifications of this parameter commonly occur using either an absolute or a relative system of measurement. In both the MKS and the CGS Systems, the basic unit of *temperature* is the **degree Kelvin** or the **degree Celsius**. The magnitudes of these two units are identical — i.e., a temperature difference between two states or conditions would have the identical numerical magnitude, whether expressed in **degrees Kelvin** or **degrees Celsius**. A temperature of 0 Kelvin, or 0 K, has been defined to be Absolute Zero; thus the Kelvin Scale is the absolute scale for these two Systems. A temperature of 0° Celsius, or 0°C, is the temperature at which water freezes; thus the Celsius Scale is the relative scale for these two Systems of Units. Please note that 273.16 K = 0°C.

BASIC PARAMETERS AND LAWS

3

In the English System, the basic unit of *temperature* is the **degree Fahrenheit** or the **degree Rankine**; as was the case with the relative and the absolute units of measure in the MKS and CGS Systems, the magnitudes of these two English System units are also identical; i.e., a temperature difference between two states or conditions would have the identical numerical magnitude, whether expressed in **degrees Fahrenheit** or **degrees Rankine**. A temperature of 0° Rankine, or 0°R, has been defined to be Absolute Zero; thus the Rankine Scale is the absolute scale for the English System. A temperature of 32° Fahrenheit, 32°F, is the temperature at which water freezes; thus the Fahrenheit Scale is the relative temperature scale for the English System. Please note that 491.67°R = 32°F.

Electrical Current

In all the Systems of Units, the basic unit of *electrical current* is the **ampere**, which has been defined to be that constant flow of electricity which, if maintained in two straight parallel conductors of infinite length, each having negligible circular cross section, and placed 1.0 meter apart in a vacuum, would produce — between these conductors and normal to the direction in which these conductors are positioned — a repulsive force equal to 2×10^{-7} newtons per meter of conductor length.

The Amount of Any Substance

In all the Systems of Units, the **mole** is the basic measure of the *amount of any substance*. The **mole** has been defined to be the precise number of elementary entities, as there are atoms in exactly 0.012 kilograms (12.0 grams) of $^{12}_{6}C$. When the **mole** is used, the specific elementary entities must be specified; however, they may be atoms, molecules, ions, electrons, protons, neutrons, other particles, or any specified groupings of such particles. In general, one **mole** of any substance will contain Avogadro's Number, N_A, of atoms, molecules, or particles of some sort. Avogadro's Number is 6.022×10^{23}.

Luminous Intensity

In all the Systems of Units, the **candela** is the unit of *luminous intensity*. The **candela** has been defined to be the luminous intensity, in a given direction, of a source that emits monochromatic radiation of a frequency equal to 5.40×10^{14} hertz, and that has a radiant intensity in that same given direction that is a fraction (specifically, 1/683) of 1.0 watt per steradian.

Supplemental Units
Plane Angle

A *plane angle* is a figure composed, in the simplest sense, of two different rays having a common endpoint. By definition these rays will always lie in a single plane.

As a supplemental *SI* Unit, the basic unit of measure for a *plane angle* is the **radian**. 1.0 **radian** is the *plane angle* formed when the tip, or end, of a rotating vector (the generator) of unit length, moving in a plane, has traveled a circular path of *length* equal to the *length* of the unit vector.

Plane angles are dimensionless quantities, since they are defined as:

$$\frac{\text{LENGTH}}{\text{LENGTH}}.$$

Clearly by this definition, there will be a total of 2π **radians** in one complete circle. In the MKS, the CGS, and the English Systems of units, *plane angles* are measured in **radians** and also frequently in **degrees**. Note that a *plane angle* of 360 **degrees** (written as 360°) = 2π **radians**, or 1.0 **radian** = 57.296°.

Solid Angle

A *solid angle* is that part of the space bounded by a moving straight line (the generator) issuing from a single point (the vertex) and moving to all points on an arbitrary closed curve. It characterizes the angle of "seeing" from which this curve is "seen".

As another supplemental *SI* Unit, the basic unit of measure for *solid angles* is the **steradian**. 1.0 **steradian** is the *solid angle* (i.e., the area) cut out of a unit sphere by the tip of a rotating unit radius vector (the generator), having the center of the sphere as its vertex and producing a *plane angle* (in *any* plane through this vertex) of 1.0 **radian** — i.e., a "cone" with its vertex angle = 1.0 **radian**.

Solid angles are dimensionless quantities. They are defined as:

$$\frac{\text{AREA}}{\text{AREA}} = \left[\frac{(\text{LENGTH})^2}{(\text{LENGTH})^2}\right].$$

In the MKS, the CGS, and the English Systems of units, *solid angles* are measured in **steradians**.

Derived Units

Area

Area is the measure of the size or extent of a surface. Its dimensions are:

$$\text{AREA} = (\text{LENGTH})^2$$

In the MKS System, *area* is measured in: **meters²**.

In the CGS System, *area* is measured in: **centimeters²**.

In the English System, *area* is measured in: **feet²**, or frequently **inches²**.

Volume

Volume is the measure of the size or extent of any three-dimensional object or region in space. Its dimensions are:

$$VOLUME = (LENGTH)^3$$

In the MKS System, *volume* is measured in: **meters³** or **liters** (with 1 meter³ = 1,000 liters).

In the CGS System, *volume* is measured in: **centimeters³**, which is numerically identical to this parameter when expressed in **milliliters**.

In the English System, *volume* is measured in: **feet³**, **inches³**, or frequently **gallons**.

Velocity or Speed

Velocity or *speed* is the distance traveled by an object during each unit interval of time. Its dimensions are:

$$VELOCITY = SPEED = \left[\frac{DISTANCE\ TRAVELED}{TIME}\right] = \left[\frac{LENGTH}{TIME}\right]$$

In the MKS System, *velocity* or *speed* is measured in:

$$\frac{meters}{second}.$$

In the CGS System, *velocity* or *speed* is measured in:

$$\frac{centimeters}{second}.$$

In the English System, *velocity* or *speed* is measured in:

$$\frac{feet}{second},\ or\ occasionally\ \frac{feet}{minute}.$$

Acceleration

Acceleration is the time rate of change of *speed* or *velocity*. Its dimensions are:

$$ACCELERATION =$$
$$\left[\frac{CHANGE\ in\ SPEED\ or\ VELOCITY}{TIME}\right] = \left[\frac{LENGTH}{TIME^2}\right]$$

In the MKS System, *acceleration* is measured in:

$$\frac{meters}{second^2}.$$

In the CGS System, *acceleration* is measured in:

$$\frac{centimeters}{second^2}.$$

In the English System, *acceleration* is measured in:

$$\frac{\text{feet}}{\text{second}^2}.$$

Force

Force is the capacity to do work or to cause change. It is a vector quantity that tends to produce an acceleration in the body on which it is acting, and to produce this acceleration in the direction of the application of that *force*. Its dimensions are:

$$\text{FORCE} = \left[(\text{MASS})(\text{ACCELERATION})\right] = \left[\frac{(\text{MASS})(\text{LENGTH})}{\text{TIME}^2}\right]$$

In the International System of Units, *force* is measured in **newtons**, which have been defined as:

$$1.0\ \textbf{newton} = 1.0\ \frac{(\text{kilogram})(\text{meter})}{\text{second}^2}.$$

In the MKS System, *force* is also measured in:

$$\textbf{newtons} = \frac{(\text{kilograms})(\text{meters})}{\text{second}^2}.$$

In the CGS System, *force* is measured in:

$$\textbf{dynes} = \frac{(\text{grams})(\text{centimeters})}{\text{second}^2}.$$

In the English System, *force* is measured in:

$$\text{pounds of force} = \frac{(\text{slug})(\text{feet})}{\text{second}^2}.$$

Pressure

Pressure is the relative magnitude of a *force* per unit *area* through which that *force* is acting. Its dimensions are:

$$\text{PRESSURE} = \left[\frac{\text{FORCE}}{\text{AREA}}\right] = \left[\frac{\text{FORCE}}{(\text{LENGTH})^2}\right] = \left[\frac{(\text{MASS})}{(\text{TIME})^2(\text{LENGTH})}\right]$$

In the International System of Units, *pressure* is measured in **pascals**, which have been defined as:

$$1.0\ \textbf{pascal} = 1.0\ \frac{\text{newton}}{\text{meter}^2}.$$

In the MKS System, *pressure* also is measured in:

$$\textbf{pascals} = \frac{\text{newtons}}{\text{meter}^2} = \frac{\text{kilograms}}{(\text{meters})(\text{second})^2}.$$

In the CGS System, *pressure* is measured in:

$$\frac{\text{dynes}}{\text{centimeter}^2} = \frac{\text{grams}}{(\text{centimeters})(\text{second})^2}.$$

In the English System, *pressure* is often measured in:

$$\frac{\text{pounds of force}}{\text{foot}^2} = \frac{\text{slugs}}{(\text{foot})(\text{second})^2},$$

or, alternatively, and even more frequently, in:

$$\frac{\text{pounds of force}}{\text{inch}^2} = \frac{(\text{slug})(\text{feet})}{(\text{inch})^2(\text{second})^2} = \frac{1}{12}\left[\frac{\text{slug}}{(\text{inch})(\text{second})^2}\right],$$

often abbreviated as **psi**.

Pressure also is frequently measured or characterized in a variety of other ways:

1. It is expressed in terms of the height of a column of some reference fluid that would be required to exert some identifiable *force* on the unit AREA on which this column of fluid is resting — i.e., (**millimeters of Mercury**), commonly written as **mm Hg**, or (**feet of Water**), commonly written as **ft H₂O**;

2. It is expressed in terms of its ratio to the average pressure of the Earth's atmosphere when measured at Mean Sea Level — i.e., (**atmospheres**), commonly written as **atms**;

3. It is expressed in terms of its ratio to the accepted *SI* unit of pressure, the **pascal** — i.e., (**torrs**), commonly written as **torrs**, of which 1.0 torr = 133.32 pascals; (**bars**), commonly written as **bars**, of which 1.0 bar = 10^5 pascals; and (**millibars**), commonly written as **mb**, of which 1.0 mb = 100 pascals.

Energy, Work, or Heat

Energy or *work* is the work that any physical system is capable of doing, as it changes from its existing state to some well-defined different state. In general, *energy* can be any of the following three types: potential, kinetic, and/or rest. *Work* is a measure of the result of applying a *force* through some identified *distance*. In the case of either *energy* or *work*, the dimensions are identical and are:

$$\text{ENERGY, WORK, or HEAT} = \left[(\text{FORCE})(\text{LENGTH})\right]$$

$$\text{ENERGY, etc.} =$$

$$\left[(MASS)(ACCELERATION)(LENGTH)\right] = \left[\frac{(MASS)(LENGTH)^2}{(TIME)^2}\right]$$

In the International System of Units, *energy* or *work* is measured in **joules**, which have been defined as:

$$1.0 \text{ joule} = 1.0 \text{ (newton)(meter).}$$

In the MKS System, *energy* or *work* is measured also in **joules**, where:

$$\textbf{joules} = (\text{newtons})(\text{meters}) = \frac{(\text{kilograms})(\text{meters})^2}{\text{second}^2}.$$

In the CGS System, *energy* or *work* is usually measured in **ergs**; however, in the context of chemical reactions, etc., it is more frequently measured in **calories**. Dimensionally, an **erg** is equal to:

$$\textbf{ergs} = (\text{dynes})(\text{centimeters}) = \frac{(\text{grams})(\text{centimeters})^2}{\text{second}^2}.$$

A **calorie** (abbreviated **cal**) has been defined to be:

$$1.0 \textbf{ calorie} = 1.0 \textbf{ cal} =$$

$$\left[\begin{array}{l}\text{the amount of heat energy required to raise the tem-}\\ \text{perature of one gram of water by one degree Kelvin}\end{array}\right].$$

In the English System, *energy* or *work* is measured in either **foot pounds** or **British Thermal Units**, where:

$$\textbf{foot pounds} = (\text{pounds of force})(\text{feet}) = \frac{(\text{slugs})(\text{feet})^2}{\text{second}^2}.$$

The **British Thermal Unit** (abbreviated **BTU**) has been defined to be:

$$1.0 \textbf{ BTU} = \left[\begin{array}{l}\text{the amount of heat energy required to raise the temperature}\\ \text{of a mass of one pound of water by one degree Fahrenheit}\end{array}\right]$$

In the microscale world (i.e., in the world of the atom, the electron, and other subatomic particles, etc.), *energy* or *work* is measured in **electron volts** (abbreviated **ev**), where this unit has been defined to be:

$$1.0 \textbf{ electron volt} = 1.0 \textbf{ ev} =$$

$$\left[\begin{array}{l}\text{the energy necessary to accelerate a single elec-}\\ \text{tron through a potential difference of one volt}\end{array}\right]$$

$$1.0 \text{ ev} \approx 1.6022 \times 10^{-19} \text{ joules}$$

Power

Power is the *time* rate of performing *work*. The dimensions of *power* are:

$$\text{POWER} = \left[\frac{\text{WORK}}{\text{TIME}}\right] = \left[\frac{(\text{FORCE})(\text{LENGTH})}{\text{TIME}}\right]$$

$$\text{POWER} =$$
$$\left[\frac{(\text{MASS})(\text{ACCELERATION})(\text{LENGTH})}{\text{TIME}}\right] = \left[\frac{(\text{MASS})(\text{LENGTH})^2}{(\text{TIME})^3}\right]$$

In the International System of Units, *power* is measured in **watts**, which has been defined as:

$$1.0 \ \textbf{watt} = 1.0 \ \frac{\text{joule}}{\text{second}}.$$

In the MKS System, *power* also is measured in **watts**, where:

$$\textbf{watts} = \frac{\text{joules}}{\text{second}} = \frac{(\text{newtons})(\text{meters})}{\text{second}} = \frac{(\text{kilograms})(\text{meters})^2}{\text{second}^3}.$$

In the CGS System, *power* is measured in:

$$\frac{\text{ergs}}{\text{second}},$$

where:

$$\frac{\text{ergs}}{\text{second}} = \frac{(\text{dynes})(\text{centimeters})}{\text{second}} = \frac{(\text{grams})(\text{centimeters})^2}{\text{second}^3}.$$

In the English System, *power* is measured in:

$$\frac{\text{foot pounds}}{\text{second}},$$

horsepower, and/or

$$\frac{\text{BTUs}}{\text{second}}.$$

Of these three general English System units, the only one that is not inherently defined by its description is the **horsepower** (abbreviated **hp**), which has been defined to be:

$$1.0 \ \textbf{hp} = 550 \ \frac{\text{foot pounds}}{\text{second}}.$$

Electric Charge

Electric charge is a measure of the quantity of electricity or charge which, in a time period of one second, passes through a section of conductor in which a constant current of one **ampere** happens to be flowing. The dimensions of *electric charge* are:

$$\text{ELECTRIC CHARGE} = \left[(\text{CURRENT})(\text{TIME})\right]$$

In both the MKS and the English System, *electric charge* is measured in **coulombs**.

In the CGS System, *electric charge* is measured in **microcoulombs**. The **coulomb** is a relatively large unit of *electric charge*. A more "natural" unit would probably be the negative charge carried by an electron (or, alternatively, the equivalently large positive charge carried by a proton). The *electric charge* carried by either of these basic atomic particles is equivalent to 1.6022×10^{-19} **coulombs**.

Electrical Potential or Potential Difference

Electrical potential is a measure of the work required to move a quantity of charge in an electrostatic field. Specifically, it is a measure of the potential energy per unit electrical charge that is characterized by any point in an electric field. The dimensions of *electrical potential* are:

$$\text{ELECTRICAL POTENTIAL} = \left[\frac{\text{WORK}}{\text{CHARGE}}\right] = \left[\frac{(\text{MASS})(\text{LENGTH})^2}{(\text{CURRENT})(\text{TIME})^3}\right]$$

In the MKS System, *electrical potential* is measured in:

$$\frac{\text{joules}}{\text{coulomb}}, \text{ or } \textbf{volts} \text{ — where: } 1.0 \textbf{ volt} = 1.0 \frac{\text{joule}}{\text{coulomb}}.$$

In the CGS System, *electrical potential* is measured in **millivolts** or **microvolts**. In the English System, *electrical potential* is also measured in **volts**, even though in this system, the unit of work is *not* the **joule**.

Capacitance

Capacitance is the ratio of (the charge on either of two conductors) to (the electrical potential difference between these two conductors). For all practical applications, it is used only to refer to the charging capacity characteristics of a "capacitor", which is a common component in many electrical circuits. The dimensions of *capacitance* are:

$$\text{CAPACITANCE} = \left[\frac{\text{CHARGE}}{\text{ELECTRICAL POTENTIAL}}\right]$$

$$= \left[\frac{(\text{CURRENT})^2(\text{TIME})^4}{(\text{MASS})(\text{LENGTH})^2}\right]$$

In the MKS System, *capacitance* is measured in **farads**, where:

$$1.0 \text{ farad} = 1.0 \ \frac{\text{coulomb}}{\text{volt}}.$$

In the CGS System, *capacitance* is measured in **microfarads** or **picofarads**.

In the English System, *capacitance* also is measured in **farads**.

Density

Density is a measure of the *mass* of anything measured per the unit *volume* that is occupied by that *mass*. The dimensions of *density* are:

$$\text{DENSITY} = \left[\frac{\text{MASS}}{\text{VOLUME}}\right] = \left[\frac{\text{MASS}}{(\text{LENGTH})^3}\right]$$

In the MKS System, *density* is measured in:

$$\frac{\text{kilograms}}{\text{meter}^3}, \text{ or occasionally in } \frac{\text{kilograms}}{\text{liter}}.$$

In the CGS System, *density* is measured in:

$$\frac{\text{grams}}{\text{centimeter}^3} = \frac{\text{grams}}{\text{milliliter}}.$$

In the English System, *density* is measured in:

$$\frac{\text{pounds of mass}}{\text{foot}^3}, \ \frac{\text{pounds of mass}}{\text{inch}^3}, \text{ or occasionally in } \frac{\text{pounds of mass}}{\text{gallon}}.$$

Concentration

Concentration is a measure of: (1) the amount or quantity of any substance per the units of *volume* that is occupied by that amount or quantity of material, (2) the mass of any substance per the units of *volume* that is occupied by that mass, (3) the ratio of the amount, quantity, or volume of any substance to the total amount, quantity, or volume of all the substances present in the overall *volume* being considered, and/or (4) the ratio of the *mass* of any substance to the total *mass* of all the substances present in the overall *volume* being considered. Let us consider each of these four *concentration* scenarios in order:

1. The first of the four listed categories of concentration is most commonly used to express the concentration of some material, substance, or chemical as a component in a solution. Its dimensions are:

$$\text{CONCENTRATION}_1 = \left[\frac{\text{MOLES}}{\text{VOLUME}}\right]$$

In the MKS System, *concentration₁* is usually measured in:

$\dfrac{\text{moles}}{\text{liter}}$; however, it can occasionally be measured in $\dfrac{\text{equivalents}}{\text{liter}}$.

In the CGS System, *concentration₁* is measured in:

$$\frac{\text{millimoles}}{\text{milliliter}} = \frac{\text{millimoles}}{\text{centimeter}^3};$$

however, it can occasionally also be measured in:

$$\frac{\text{milliequivalents}}{\text{milliliter}} = \frac{\text{milliequivalents}}{\text{centimeter}^3}.$$

The English System does not have a commonly used unit of *concentration₁*.

2. The second of the four listed categories of *concentration* is most commonly used to express the concentration of a particulate, an aerosol, a mist, a dust, or a fume in the air. Its dimensions are:

$$\text{CONCENTRATION}_2 = \left[\frac{\text{MASS}}{\text{VOLUME}}\right]$$

In the MKS System, *concentration₂* is measured in:

$$\frac{\text{milligrams}}{\text{meter}^3}.$$

In the CGS System, *concentration₂* is measured in:

$$\frac{\text{micrograms}}{\text{centimeter}^3} = \frac{\text{micrograms}}{\text{milliliter}}.$$

The English System does not have a commonly used unit of *concentration₂*.

3. The third category of *concentration* is most commonly used to express the concentration of some vapor or gas in the air. This category of *concentration* is considered to be *volume-based concentration*. Its dimensions are:

$$\text{CONCENTRATION}_3 = \left[\frac{\text{MOLES}}{\text{MOLE}}\right],$$

or more commonly

$$\left[\frac{\text{MOLES}}{\text{MILLION MOLES}}\right]$$

In all of the systems of units, namely: the MKS, the CGS, and the English Systems, *concentration₃* is measured in:

$$\frac{\text{parts}}{\text{million parts}}, \text{ or, simply, } \textbf{ppm(vol)}.$$

4. The fourth and final category of *concentration* is most commonly used to express concentrations of some mixture of materials, or solid sample of some sort. This category of *concentration* is considered to be *mass-based concentration*. Its dimensions are:

$$\text{CONCENTRATION}_4 = \left[\frac{\text{MASS}}{\text{TOTAL MASS}} \right]$$

In the MKS and the CGS Systems, *concentration$_4$* is measured in:

$$\frac{\text{milligrams}}{\text{kilogram}}, \text{ or, more simply, } \textbf{ppm(mass)}.$$

The English System does not have a commonly used unit of *concentration$_4$*.

Luminous Flux

Luminous flux is the product of the *luminous intensity* and the *solid angle* over which the illumination being considered is being emitted. *Luminous flux* is measured in **lumens**, and its dimensions are:

$$\text{LUMINOUS FLUX} = [(\text{LUMINOUS INTENSITY})(\text{SOLID ANGLE})]$$

By definition, a point light source radiating light energy in all directions and having a *luminous intensity* of 1.0 **candela** will produce a total LUMINOUS FLUX of 4π **lumens**.

The basic unit of *luminous flux* is the same in all systems of units, and is the **lumen**.

Frequency

Frequency is a measure of the number of cycles per unit *time* for any periodically repeating physical phenomenon — i.e., a sound, or an electromagnetic wave, each of which exhibits a form of simple harmonic motion. The dimensions of *frequency* are:

$$\text{FREQUENCY} = \frac{\text{CYCLES}}{\text{TIME}} = \frac{1}{\text{TIME}}$$

The basic unit of measure for *frequency* is the same in all systems of units, and is the **hertz**, which has been defined to be:

$$1.0 \textbf{ hertz} = \frac{\text{one complete cycle}}{\text{second}}$$

Radioactive Activity

The basic event that characterizes any radioactive element or nuclide is the transformation or decay of its nucleus into the nucleus of some other species. The number of such transformations per unit time is known as the *radioactive activity* of the element or nuclide being considered. The basic unit of *radioactive activity* is the **becquerel**, which is defined to be one transformation per second. Its dimensions are:

$$\text{RADIOACTIVE ACTIVITY} = \frac{\text{DECAY EVENTS}}{\text{TIME}} = \frac{1}{\text{TIME}}$$

Its dimensions are, at least from the perspective of pure dimensionality, identical to those of *frequency*, namely,

$$\frac{1}{\text{TIME}};$$

however, there is a clear difference as the seemingly dimensionless "numerator" for each of these two parameters is totally different.

The basic *SI* unit of *radioactive activity* is the **becquerel**.

This is also the basic unit of *radioactive activity* for the MKS, the CGS, and the English Systems. In each of these systems, additional widely used units of *radioactive activity* are: the **curie**, the **millicurie**, the **microcurie**, and the **picocurie**, 1.0 **curie** = 2.22×10^{12} **becquerels**. The **becquerel** is a *very small* unit, while the **curie** is *extremely large*. Because of this, one most commonly encounters the **millicurie** (1/1,000 of **curie**), the **microcurie** (1/1,000 of a **millicurie**) and the **picocurie** (1/1,000 of a **microcurie**).

Absorbed Radiation Dose

The basic quantity that is used to characterize the amount of energy (in the form of some type of radiation) that has been imparted to matter is the *absorbed radiation dose*. The *absorbed radiation dose* is the ratio of (the radioactive energy imparted to the matter in that region) to (the mass of the matter in that region). The dimensions of *absorbed radiation dose* are:

$$\text{ABSORBED RADIATION DOSE} = \frac{\text{ENERGY}}{\text{MASS}} = \left[\frac{(\text{LENGTH})^2}{(\text{TIME})^2} \right]$$

The basic *SI* unit of *absorbed radiation dose* is the **gray**.

This is also the basic unit of *absorbed radiation dose* for each of the other systems of units, namely, the MKS, the CGS, and the English Systems. In these latter systems additional very widely used units of *absorbed radiation dose* are the **rad**, the **millirad**, and the **microrad**. By definition:

$$1.0 \text{ gray} = 1.0 \frac{\text{joule}}{\text{kilogram}} = 100 \text{ rads.}$$

Therefore,

$$1.0 \textbf{ rad} = 100 \frac{\text{ergs}}{\text{gram}}.$$

The energy of most radioactive particles is most frequently expressed in millions of electron volts (MeV); therefore, it is useful to quantify these *absorbed radiation doses* in terms of this unit of energy, thus:

$$1.0 \textbf{ rad} = 6.24 \times 10^7 \frac{\text{MeV}}{\text{gram}}, \text{ and } 1.0 \textbf{ millirad} = 62,400 \frac{\text{MeV}}{\text{gram}}.$$

In general, harmful levels of *absorbed radiation dose* are expressed in **rads**, whereas the **millirad** (often abbreviated **mrad**) is most frequently used to specify permissible dose rates. Occupational radioactivity exposures should *never* exceed a few **mrads** per hour.

Radiation Dose Equivalent
or Radiation Dose Equivalent Index

The basic quantity used in radiation protection, as it applies to humans, is the *radiation dose equivalent*. Injuries produced as a result of an individual's exposure to radioactivity depend not only on the total of the radioactive energy imparted to that person, but also on the nature of the radiation that produced the exposure. The dimensions of the *radiation dose equivalent* are:

$$\text{RADIATION DOSE EQUIVALENT} = \frac{\text{ENERGY}}{\text{MASS}} = \left[\frac{(\text{LENGTH})^2}{(\text{TIME})^2} \right]$$

> **Note:** *The* absorbed radiation dose *and the* radiation dose equivalent *are dimensionally equal. We must use this additional parameter (i.e., the Radiation Dose Equivalent, etc.) because of differences in the harmful effects one type of radiation can produce when compared to other types. As an example, a 1.0* **mrad** *absorbed radiation dose from neutrons would produce a far greater injury to an exposed individual than would the same* absorbed radiation dose *from beta particles. Because of this variability in harmful effects, it is clear that if we use only the simple* absorbed radiation dose *parameter, we will fail to provide a completely accurate and unambiguous picture when evaluating the injuries that result from exposure to various different types of radiation.*

Although, as stated above, the dimensionalities of *absorbed radiation dose* and *radiation dose equivalent* are identical, for the sake of clarity, there is a different unit for *radiation dose equivalent* — this unit is the **sievert**. In the MKS, the CGS, and the English Systems, an additional unit of *radiation dose equivalent* is also in wide use. This unit is the **rem**.

The "effectiveness" in producing injury for any one category of radiation relative to any other can vary widely depending on a number of factors and circumstances. Included among these are:

1. the nature of the biological material exposed,
2. the length of time during which the exposure occurred,
3. the type of effect being considered, and
4. the ability of the exposed body to repair the injuries produced by the radiation.

Therefore, when considering aspects of radiation protection, an occupational safety and health professional must evaluate the manner in which various different forms of ionizing radiation impart energy to the tissue onto which it has fallen.

The physical measure that is used for evaluating the "effectiveness" of equal *absorbed radiation doses* from different types of radiation — insofar as the ability of the radiation to produce injury — is the Linear Energy Transfer, or *LET*. The greater the *LET* for any type of radiation, the greater will be the injury produced for any given *absorbed radiation dose*. The quantitative factor that expresses the relative "effectiveness" of a given type of radiation, based on its *LET*, is the *Quality Factor*. Numeric values of these *Quality Factors*, as functions of the radiation type and energy, have been developed from animal experimental data, and are given in **Table 1**.

Table 1.   Approximations to a Variety of Different *Quality Factors*

Particle Type	Energy Related Parameters	Quality Factor
X- and Gamma Rays	any	1
Beta Particles	any	1
Heavy Ionizing Particles	LET ≤ 35 MeV/cm	2
	35 MeV/cm < LET < 70 MeV/cm	2 to 4
	70 MeV/cm < LET < 230 MeV/cm	4 to 10
	230 MeV/cm < LET < 530 MeV/cm	10 to 20
	530 MeV/cm < LET < 1,750 MeV/cm	20 to 40
	LET ≥ 1,750 MeV/cm	40
Neutrons	thermal	5
	0.0001 MeV	4
	0.01 MeV	5
	0.1 MeV	15
	0.5 MeV	22
	1.0 MeV	22
	10 MeV	13
Protons	≤ 10 MeV	20
Alpha Particles	≤ 10 MeV	20
Other Multiply Charged Particles	≤ 10 MeV	20

Although the issue is even more complex than has been indicated to this point, this discussion will go no further. When the *absorbed radiation dose* is evaluated for the purposes of radiation protection, this dose number must be multiplied by the appropriate *Quality Factor* in order to provide an accurate *radiation dose equivalent*. Specifically, the following relationships provide the method for obtaining accurate values of the *radiation dose equivalent* for any measured *absorbed radiation dose*, each as a function of the specific units of these two similar parameters:

$$\text{rems} = \left[(\text{rads})(\text{QF})\right], \text{ and}$$

$$\text{sieverts} = \left[(\text{grays})(\text{QF})\right].$$

As stated above, a tabulation listing some of these *Quality Factors* — from data published in 1987 by the National Council on Radiation Protection — is shown in **Table 1**.

Atmospheric Standards
Standard Temperature and Pressure

Standard Temperature and Pressure (usually abbreviated **STP**) is the designation given to an ambient condition in which the Barometric Pressure (**P**) and the Ambient Temperature (**T** or **t**) are, respectively:

Barometric Pressure (P)		Ambient Temperature (T or t)
1 atmosphere, or		0°C, or
760 mm Hg, or		32°F, or
14.70 psia, or	and	273.16 K, or
0.00 psig, or		491.67°R
1,013.25 millibars, or		
760 Torr		

Normal Temperature and Pressure

Normal Temperature and Pressure (usually abbreviated **NTP**) is the designation given to an ambient condition in which the Barometric Pressure (**P**) and the Ambient Temperature (**T** or **t**) are, respectively:

Barometric Pressure (P)		Ambient Temperature (T or t)
1 atmosphere, or		25°C, or
760 mm Hg, or		77°F, or
14.70 psia, or	and	298.16 K, or
0.00 psig, or		536.67°R
1,013.25 millibars, or		
760 Torr		

Ventilation-Based Standard Air

The conditions that are characteristic of ventilation-based **Standard Air** differ slightly from those of either Standard Temperature and Pressure or Normal Temperature and Pressure listed above. **Standard Air** is the designation given to ambient air for which the following *three specific conditions must* be met, namely:

Barometric Pressure(P)	Ambient Temperature (T or t)	Humidity Level
1.00 atmosphere, or	21.1°C, or	Relative Humidity = 0.0%
760 mm Hg, or	70°F, or	Absolute Humidity = 0.0%
29.92 inches of Hg, or	294.26 K, or	Water Conc. = 0.0 ppm(vol)
14.70 psia, or	529.67°R	
0.00 psig, or		i.e., Standard Air is *dry*
1,013.25 millibars, or		
760 Torr		

Standard Air also is regarded as possessing the following characteristics of density and specific heat:

$$\text{Density of } \textbf{Standard Air} = \rho_{\text{Standard Air}} = 0.075 \text{ lbs/ft}^3 = 1.201 \times 10^{-3} \text{ gm/cm}^3$$

$$\text{Specific Heat of } \textbf{Standard Air} = C_{P-\text{Standard Air}} = 0.24 \text{ BTU/lb/°F.}$$

Metric Prefixes (for use with *SI* Units)

	"Expanding" Prefixes			"Diminishing" Prefixes	
Prefix	*Abbreviation*	*Multiplier*	*Prefix*	*Abbreviation*	*Multiplier*
deca-	da	10^1	deci-	d	10^{-1}
hecto-	h	10^2	centi-	c	10^{-2}
kilo-	k	10^3	milli-	m	10^{-3}
mega-	M	10^6	micro-	μ	10^{-6}
giga-	G	10^9	nano-	n	10^{-9}
tera-	T	10^{12}	pico-	p	10^{-12}
peta-	P	10^{15}	femto-	f	10^{-15}
exa-	E	10^{18}	atto-	a	10^{-18}
zetta-	Z	10^{21}	zepto-	z	10^{-21}
yotta-	Y	10^{24}	yocto-	y	10^{-24}

Relevant Formulae and Relationships
Temperature Conversions

As stated earlier, temperatures can be expressed in relative or absolute units, and in the English or the Metric Systems (i.e., CGS, MKS, and/or *SI* Systems). To convert from one form of temperature units to any other, the following four relationships should be used: Equation 1.1, to convert from relative to absolute Metric System units; Equation 1.2, to convert from relative to absolute English System units; Equation 1.3, to convert from relative Metric to relative English System units; and Equation 1.4, to convert temperature changes in any Metric System unit (relative or absolute) to the corresponding temperature change in English System units (relative or absolute).

Equation 1.1

$$t_{Metric} + 273.16 = T_{Metric}$$

Where: t_{Metric} =   the temperature, expressed in relative Metric units — i.e., degrees Celsius (°C);

         T_{Metric} =   the temperature, expressed in absolute Metric units — i.e., degrees Kelvin (K).

Equation 1.2

$$t_{English} + 459.67° = T_{English}$$

Where: $t_{English}$ =   the temperature, expressed in relative English units — i.e., degrees Fahrenheit (°F);

         $T_{English}$ =   the temperature, expressed in absolute English units — i.e., degrees Rankine (°R).

Equation 1.3

$$t_{Metric} = \frac{5}{9}\left[t_{English} - 32°\right]$$

Where: $t_{English}$ and t_{Metric} are as defined above on this page for the two previous equations.

Equation 1.4

$$\Delta t_{Metric} = \frac{5}{9}\Delta t_{English}$$

and/or

$$\Delta T_{Metric} = \frac{5}{9}\Delta T_{English}$$

Where: Δt_{Metric} =   the temperature change in degrees Celsius (°C);

$\Delta t_{English}$ = the temperature change in degrees Fahrenheit (°F);

ΔT_{Metric} = the temperature change in degrees Kelvin (K); and

$\Delta T_{English}$ = the temperature change in degrees Rankine (°R).

The Standard Gas Laws

The following Formulae make up the five Standard Gas Laws, which are, in the order in which they will be presented and discussed: Boyle's Law (Equation 1.5); Charles' Law (Equation 1.6); Gay-Lussac's Law (Equation 1.7); the General Gas Law (Equation 1.8); and the Ideal or Perfect Gas Law (Equation 1.9).

Equation 1.5

The following relationship, Equation 1.5, is Boyle's Law, which describes how the Pressure and Volume of a gas vary *under conditions of constant temperature.*

$$P_1V_1 = P_2V_2$$

Where: P_1 = the Pressure of a gas @ Time 1, measured in some suitable pressure units;

V_1 = the Volume of that same gas @ Time 1, measured in some suitable volumetric units;

P_2 = the Pressure of that same gas @ Time 2, measured in the *same* pressure units as P_1 above; and

V_2 = the Volume of that same gas @ Time 2, measured in the *same* volumetric units as V_1 above.

Equation 1.6

The following relationship, Equation 1.6, is Charles' Law, which describes how the Volume and the Absolute Temperature of a gas vary *under conditions of constant pressure.*

$$\frac{V_1}{T_1} = \frac{V_2}{T_2}$$

Where: V_1 and V_2 are the Volumes of the gas of interest at each of its two states, with this term as was defined for Equation 1.5;

T_1 =   the Absolute Temperature of a gas @ Time 1, measured in either K or °R; and

T_2 =   the Absolute Temperature of the same gas @ Time 2, measured in the *same* Absolute Temperature units as T_1.

Equation 1.7

The following relationship, Equation 1.7, is Gay-Lussac's Law, which describes how the Pressure and Temperature of a gas vary *under conditions of constant volume*.

$$\frac{P_1}{T_1} = \frac{P_2}{T_2}$$

Where: P_1 and P_2 are the Pressures of the gas of interest at each of its two states, with this term as was defined for Equation 1.5; and

T_1 and T_2 are the Absolute Temperatures of the gas of interest at each of its two states, with this term as was defined for Equation 1.6.

Equation 1.8

The following formula, Equation 1.8, is the General Gas Law, which is the more generalized relationship involving changes in *any* of the basic measurable characteristics of *any* gas. This relationship permits the determination of the value of *any* of the three basic characteristics of the gaseous material being evaluated — namely: its Pressure, its Temperature, and/or its Volume — any one of which might have changed as a result of changes in either one or both of the other two characteristics.

$$\frac{P_1V_1}{T_1} = \frac{P_2V_2}{T_2}$$

Where: P_1 and P_2 are the Pressures of the gas of interest at each of its two states, with this term as was defined for Equation 1.5;

V_1 and V_2 are the Volumes of the gas of interest at each of its two states, with this term as was defined for Equations 1.5 and 1.6; and

T_1 and T_2 are the Absolute Temperatures of the gas of interest at each of its two states, with this term as defined for Equations 1.6 and 1.7.

Equation 1.9

The following relationship, Equation 1.9, is the Ideal Gas Law (it is also frequently called the Perfect Gas Law). This law is one of the most commonly used Equations of State. Like the immediately preceding formula, this one provides the necessary relationship for determining the value of *any* of the measurable characteristics of a gas — namely, again: its Pressure, its Temperature, and/or its Volume; however, it *does not require* that one know these characteristics at some alternative state or condition. Unlike Equation 1.8, it *does require* that the *quantity* of the gas involved in the determination be known (i.e., the Number of Moles involved, or the weight and identity of the gas involved, etc.).

$$PV = nRT$$

Where: **P** $=$ the Pressure of the gas involved, in some appropriate units;

         V $=$ the Volume of the gas involved, in some appropriate units;

         n $=$ the number of moles of the gas involved (for more detailed information on this "number of moles" factor, see the discussion under Equations 1.10 and 1.11);

         R $=$ the Universal Gas Constant, in units consistent with those used for the Pressure, **P**, the Volume, **V**, and the temperature, **T**:

$$= \quad 0.0821 \, \frac{(\text{liter})(\text{atmospheres})}{(\text{K})(\text{moles})};$$

$$= \quad 62.36 \, \frac{(\text{liter})(\text{mm Hg})}{(\text{K})(\text{mole})};$$

$$= \quad 1.631 \, \frac{(\text{feet})^3 (\text{millibars})}{(°\text{R})(\text{mole})};$$

$$= \left[\begin{array}{l} \text{The Universal Gas Constant can be ex-} \\ \text{pressed in a wide variety of units so as} \\ \text{to provide a suitable proportionality for} \\ \text{any application of the Ideal Gas Law.} \end{array} \right]$$

T   =   the Absolute Temperature of the gas involved, in some appropriate units of absolute temperature.

The Mole

Equation 1.10

The following Equation, 1.10, involves the Mass-Based relationship that defines the **mole**. Specifically, a mole (which for this relationship is equivalent to a *gram mole* or a *gram molecular weight*) of any compound or chemical is that quantity, which when weighed will have a weight, *in grams*, that is *numerically equivalent* to that compound's or chemical's molecular weight, when that molecular weight is expressed in Atomic Mass Units (amu). Analogously, we can also define the pound mole of any compound or chemical as that quantity, which when weighed will have a weight, *in pounds*, that is numerically equivalent to that compound's or chemical's molecular weight, again when this molecular weight is expressed in Atomic Mass Units (amu). The pound mole is only very rarely ever used, particularly when compared to the *SI* mole — thus, when a reference identifies only "a mole," it can always be assumed that it is the *SI* mole that is involved. A pound mole has a mass that is 453.59 times greater than that of an *SI* mole.

$$n = \frac{m}{MW}$$

Where: **n**   =   the number of moles of the chemical or the material being evaluated;

     m   =   the mass, measured in grams, of the chemical or the material being evaluated; and

     MW   =   the molecular weight of the chemical or the material being evaluated, expressed in Atomic Mass Units (amu).

Equation 1.11

Equation 1.11 involves the numerical count-based relationship that also defines the mole. A mole (i.e., a gram mole) of *any* compound or chemical *always will contain* Avogadro's Number, usually designated as N_A, which equals — numerically — 6.022×10^{23} particles (i.e., molecules, atoms, ions, etc.).

$$n = \frac{Q}{N_A} = \frac{Q}{6.022 \times 10^{23}}$$

Where: **n**   =   the number of moles, as defined in the discussion of Equation 1.10;

               Q   =   the actual count or number of particles (i.e., molecules, atoms, ions, etc.) of the chemical or material being evaluated; and

               N_A   =   Avogadro's Number = 6.022×10^{23}

Equations 1.12 and 1.13

The following two equations, 1.12 and 1.13, incorporate a third relationship that can be used to identify the quantity of any substance. Again, these relationships also serve to define the mole. In these cases the relationships involved are volumetric rather than weight based. It is known that a mole of *any* compound or chemical *that exists as a gas or vapor* will occupy a very specific volume under the prevailing conditions of pressure and temperature at which the evaluation is being made. Under STP Conditions, the Molar Volume$_{STP}$ = 22.414 liters; under NTP Conditions, the Molar Volume$_{NTP}$ = 24.465 liters.

Equation 1.12

$$n = \frac{V_{sample}}{V_{molar-STP}} = \frac{V_{sample}}{22.414} \text{ @ } \mathbf{STP} \text{ Conditions}$$

Where: **n**   =   the number of moles, as defined in Equation 1.10;

            V_{sample}   =   the volume, expressed in liters, of the sample of gaseous material being evaluated — the sample for which the specific quantity or amount of material present must be determined; and

            $V_{molar-STP}$ =   the Molar Volume$_{STP}$, expressed in liters, under conditions Standard Temperature and Pressure — under these specific conditions of temperature and pressure, this Molar Volume is **22.414 liters**.

Equation 1.13

$$n = \frac{V_{sample}}{V_{molar-NTP}} = \frac{V_{sample}}{24.465} \text{ @ } \textbf{NTP} \text{ Conditions}$$

Where: n =   the number of moles, as defined in Equation 1.10;

V_{sample} =   the volume, expressed in liters, as defined above on this, as well as under equation 1.12; and

$V_{molar-NTP}$ =   the Molar Volume$_{NTP}$, expressed in liters, under conditions Normal Temperature and **Pressure** — under these specific conditions of temperature and pressure, this Molar Volume is **24.465 liters**.

Other Gas Laws

Equation 1.14

The following Equation, 1.14, provides a vehicle for relating the Density of *any* gas to its Pressure and Absolute Temperature characteristics.

$$\frac{\rho_1 T_1}{P_1} = \frac{\rho_2 T_2}{P_2}$$

Where: P_1 and P_2 are the Pressures of the gas of interest at each of its two states, as was defined for Equation 1.5;

T_1 and T_2 are the Absolute Temperatures of the gas of interest at each of its two states, as was defined for Equation 1.6;

ρ_1 =   the Density of the gas @ Time 1, measured commonly, in units of

$$\frac{grams}{cm^3}; \text{ and}$$

ρ_2 =   the Density of the gas @ Time 2, also measured in units of

$$\frac{grams}{cm^3},$$

but most importantly, in the same units as was the case for ρ_1.

Equations 1.15 and 1.16

The following two equations, 1.15 and 1.16, are expressions of Dalton's Law of Partial Pressures; in general, this Law simply states that the total pressure exerted by the mixture of different gases in any volume will be, and is, equal to the *sum of the partial pressures* of each individual component in the overall gas mixture.

Dalton's Law of Partial Pressures also states that each individual component in any gas mixture exerts its own individual partial pressure, in the *exact ratio* as its mole fraction in that mixture.

Equation 1.15

$$P_{total} = \sum_{i=1}^{n} P_i = P_1 + P_2 + \ldots + P_n$$

Where:　　$\mathbf{P_{total}}$　　=　　the Total Pressure of some volume or system containing a total of "**n**" different identifiable gaseous components, with all pressures measured in some suitable and consistent units; and

　　　　　　$\mathbf{P_i}$　　=　　the Partial Pressure of the **i**th member of the "**n**" total gaseous components of the entire mixture being considered, again with all pressures measured in some suitable and consistent units.

The second portion of the overall statement of Dalton's Law of Partial Pressures involves the previously mentioned relationship of the partial pressure of any component to its mole fraction in the entire mixture. This relationship also provides a way to determine the volume-based Concentration of any individual gaseous component in a mixture of several other gases OR in the ambient air. This relationship can be usefully expressed in a variety of different forms:

Equation 1.16

$$P_i = P_{total} m_i \text{ or } P_i = \frac{P_{total} C_i}{1,000,000} \text{ or } C_i = \frac{[1,000,000] P_i}{P_{total}}$$

Where:　　$\mathbf{P_i}$　　=　　the Partial Pressure of the **i**th member of the "**n**" total gaseous components of the entire mixture being considered, as defined for Equation 1.15;

$\mathbf{P}_{total}$ = the Total Pressure of some volume or system containing a total of "**n**" different identifiable gaseous components, as defined for Equation 1.15;

$\mathbf{m}_i$ = the Mole Fraction of the ith component in the entire mixture of "**n**" total gaseous components; and

$\mathbf{C}_i$ = the Concentration of the ith component in the entire mixture of "**n**" total gaseous components, expressed in ppm(vol).

Equations 17 and 18

The following two relationships, Equations 1.17 and 1.18, are known as Raoult's Law. This Law involves the relationship between the partial vapor pressure of each of the components in a solution containing two or more volatile components — i.e., the partial pressure of each component in the mixed vapor phase that exists in equilibrium with the solution — *and* the solution mole fraction of each component (*not the vapor phase mole fraction!*). The basic relationship utilizes the Vapor Pressure each component would exert if it was alone in the pure liquid state at the temperature of the solution. The first of these two relationships involves a consideration of just one single component member of the solution being considered.

Equation 1.17

$$PVP_i = m_i VP_i \quad \text{for the ith component in the solution}$$

Where: $\mathbf{PVP}_i$ = the Partial Vapor Pressure — in the vapor space above the solution — produced by the ith component of a solution containing "**n**" different components, with this pressure measured in some suitable and consistent units of pressure; and

$\mathbf{m}_i$ = the Solution Mole Fraction of the ith component in a solution containing a total of "**n**" different components; and

$\mathbf{VP}_i$ = the **Vapor Pressure** of the "pure" ith component of the solution, always listed as a function of its temperature, with this pressure measured in some suitable and consistent units of pressure.

Equation 1.18

This second expression involves considering the multicomponent solution as a whole.

$$TVP_{solution} = \sum_{i=1}^{n} PVP_i = \sum_{i=1}^{n} m_i VP_i = m_1 VP_1 + m_2 VP_2 + \cdots + m_n VP_n$$

Where: **TVP$_{solution}$**   =   the Total Vapor Pressure for all the volatile components in the solution, with this pressure also measured in some suitable and consistent units of pressure; and

PVP$_i$, m$_i$, and VP$_i$ are as defined for Equation 1.17.

Equation 1.19

The following relationship, Equation 1.19, constitutes the definition of Absorbance. Absorbance is the most commonly used measure of the decrease in intensity of any beam of electromagnetic radiation, so long as this decrease has been produced by the absorption of some fraction of the energy of that beam by the components of the matrix through which it is passing. Absorbance is a dimensionless parameter, but its quantitative measure is frequently referred to and identified as "Absorbance Units".

$$A = \log \frac{I_0}{I}$$

Where: **A**   =   Absorbance produced in a beam of electromagnetic energy produced by the absorption of some fraction of that beam by components of the matrix through which that beam has passed, measured in Absorbance Units or AUs.

I$_0$ =   the Intensity, or Total Energy, of the beam as it enters the matrix through which it will be passing, and

I   =   the Intensity, or Total Energy, of the beam as it exits the matrix through which it has passed.

Settling Velocity of Various Types of Particulates
Suspended in Air

Equations 1.20 and 1.21

These two Equations, 1.20 and 1.21, are, in order: (1) for Equation 1.20, the rigorous statement of Stoke's Law, which relates the terminal, or settling velocity, *in air* of any suspended particle (i.e., dusts, fumes, mists, etc.) to the combination of that particle's effective diameter and density, and (2) for Equation 1.21, the most useful form of this equation, simplified from its more precise cousin by ignoring the contribution of the numerically very small density of air. In *every case*, the density of the settling particle *must be greater* than the density of the air — which will virtually always be less than 0.0013 grams/cm³ — in order for the particle to settle at all. Particles less dense than air are *extremely rare*, and because of their low density, they will *never* settle at all. At the other extreme, particles with an effective diameter *greater* than 50 to 60 microns will simply "fall" to earth, rather than "settle"; thus this approximation to Stoke's Law (Equation 1.20) *never* should be used in any case where the effective diameter of the particle being considered is at or above this size range.

Equation 1.20

$$V_s = 0.003\left[\rho_{particle} - \rho_{air}\right]d^2_{particle}$$

Where: V_s = the Settling Velocity in air of the particle being considered, measured in cm/sec;

$\rho_{particle}$ = the Density of the particle being considered, measured in gms/cm³;

$\rho_{air\text{-}STP}$ = the Density of air under STP conditions, identified as 0.00129 gms/cm³;

$\rho_{air\text{-}NTP}$ = the Density of air under NTP conditions, identified as 0.00118 gms/cm³; and

$d_{particle}$ = the Effective Diameter of the particle, measured in microns, (μ).

Equation 1.21

$$V_s = 0.003\left[\rho_{particle}\right]\left[d^2_{particle}\right]$$

Where: V_s =   the Settling Velocity in air of the
                                      particle being considered, measured
                                      in cm/sec;

                $\rho_{particle}$ =   the Density of the particle being
                                      considered, as defined for Equation
                                      1.20; and

                $d_{particle}$ =   the Effective Diameter of the particle,
                                      also as defined for Equation 1.20.

In general, as stated above, the density of most particulate matter will
always be *far greater* than the density of air in which the particles being
considered are suspended; therefore, we can simply ignore the contribution
to this quantitative relationship, represented by Equation 1.20, that results
from the subtraction of the extremely small density of air from the very large
density of the particulate matter being considered. Equation 1.21 is the result
of this simplification.

Chapter 2
Standards and Calibrations

This chapter will discuss reference methods, procedures, and standards against which all field measurements must be compared. The validity of any measurement will depend on the accuracy of the method, procedure, technique, and instrumentation used to make it. Factors such as the precision, accuracy, and/or repeatability of any analytical effort completed outside of the laboratory can be and frequently are called into question. The individual who has made a challenged measurement in the field or in the lab must be able to document the relationship between the result he or she has reported and an appropriate, accepted, and well-established standard.

Relevant Definitions
Primary Standard

A standard for any measurable parameter (i.e., time, length, mass, etc.) that is maintained by any of the international or national standards agencies, most commonly by either the United States National Institute of Standards and Technology (NIST), in Washington, DC — formerly known as the United States National Bureau of Standards (NBS) — or the International Organization for Standardization (ISO), in Geneva, Switzerland.

Secondary Standard

A standard for any measurable parameter (i.e., temperature, volume, etc.) that is maintained by any commercial, military, or other organization — excluding any of those groups referenced above, i.e., groups that maintain Primary Standards. A Secondary Standard will have been thoroughly documented as to the fact of its having been directly referenced against an appropriate and applicable Primary Standard. Common Secondary Standards include such things as balance weights, atomic clocks, etc.

Standard Reference Material

A Standard Reference Material — often abbreviated as SRM — is any material, item, etc. for which one or more important characteristics (i.e., the specific makeup of a mixture such as Arizona Road Dust, the leak rate of a gas permeation device, the purity of a radioactive chemical, the precision and accuracy of a liquid-in-glass thermometer, etc.) have been certified by well-documented procedures to be traceable to some specific Primary Standard. Standard Reference Materials can be obtained from the National Institute of Standards and Technology, or any commercial supplier. In every case the SRM will have had the specific characteristic of interest to its purchaser certified as being traceable to the appropriate Primary Standard.

Calibration

Calibration is a process whereby the operation or response of any analytical method, procedure, instrument, etc. is referenced against some standard — most likely either a Secondary Standard directly, or some mechanism that incorporates a Secondary Standard. As an example, let us consider a situation wherein the actual response of a gas analyzer that has been designed to measure some specific analyte is unknown. Such an analyzer might be challenged with a number of known concentrations of the vapor of interest — with the known gas concentrations having been generated by a system that employs a Secondary Standard as its vapor source. This type of process, known as a Calibration, will document the previously unknown relationship between the analyzer response and the specific vapor concentrations that have produced each response. Such a Calibration would result in a curve or plot showing the analyzer output vs. vapor concentration.

Calibration Check

A Calibration Check is a simple process where a previously calibrated method, procedure, instrument, etc. is challenged, most commonly with a "Zero" and a single "Non-Zero" calibration standard — this latter one again most likely either a Secondary Standard directly, or some mechanism that incorporates a Secondary Standard. Such a "Non-Zero" challenge is frequently referred to as a "Span Check"; it serves primarily to confirm that the system in question is working properly. A Calibration Check can also involve multipoint ("Zero" and multiple "Non-Zero") challenges designed to confirm that a system in question is responding properly over its entire designed operating range.

Sensitivity

Sensitivity is a measure of the smallest value of any parameter that is to be monitored that can be unequivocally measured by the system being considered. It is a function of the inherent noise that is present in any analytical system. Sensitivity is almost always defined and/or specified by a manufacturer as some multiple (usually in the range of 2X to 4X) of the zero level noise of the system being considered. As an example, if some type of analytical system were to produce a steady ± 0.1 mv output when it is being exposed to a zero level of whatever material it has been designed to measure, then one might specify the Sensitivity of this system to that analyte level that would produce a 0.2 to 0.4 mv output response.

Selectivity

Selectivity is the capability of any analytical system to provide accurate answers to specific analytical problems even in the presence of factors that might potentially interfere with the overall analytical process. Selectivity is most easily understood by considering a typical example; in this case we will consider sound measurements. Suppose we are dealing with an Octave

Band Analyzer that has been set up to provide equivalent sound pressure levels for the 1,000 Hz Octave Band. Suppose further that the sounds being monitored include all frequencies from 20 to 20,000 Hz. The Selectivity of this analytical tool would be its ability to provide accurate measurements of the 1,000 Hz Octave Band while simultaneously rejecting the contributions of any other segment of the entire noise spectrum to which it was exposed.

Repeatability

Repeatability is the ability of an analytical system to deliver consistently identical results to specific identical analytical challenges independent of any other factors. Specifically, an analytical system can be said to be repeatable if it provides the same result ($\pm$ a small percentage of this result) when challenged with a known level of the material for which this system was designed to monitor.

Although the following listing is not necessarily complete, a repeatable system would have to perform as listed above under any or all of the following conditions: (1) different operators; (2) different times of day; (3) an "old" system vs. a "new" one, etc.

Timeliness

The Timeliness of any measurement is related to the interval of time between the introduction of a sample to an analytical system, and the time required for that system to provide the desired result. Systems are classified into one of the three following groupings, each as a function of this specific time interval, or delay time, to provide an analytical answer. These groupings are:

1. *Instantaneous or Real-Time*   Any system that provides its analytical output at the same time as it is presented with the sample. Instantaneous or Real-Time systems are the only types that are capable of determining true Exposure Limit Ceiling Values.

2. *Slow*   Any system that has a delay interval between a few seconds and 30 minutes would be called a Slow system. A gas chromatograph would fall into this category.

3. *Very Slow*   Any system that will typically require days to be able to provide its answer. Dosimeters of all types tend to fall into this grouping.

Accuracy

The Accuracy of any measurement will simply be the value that has been specified by the manufacturer of the instrument involved. For most analytical instruments, the manufacturers will have identified the specific unit's Accuracy as a percentage of its full scale reading. As an example, a Carbon Dioxide Analyzer that has been set up to operate in the range 0 to 2,000 ppm (0 to 0.2%) will typically have an Accuracy Specification of ± 10% of its full scale reading, or ± 200 ppm (200 ppm = 10% of 2,000 ppm). Although it is not yet common, some manufacturers now specify Accuracies for their instruments in terms of a combination of: (1) a percentage of the analyzer's full scale reading and (2) a percentage of the actual reading, whichever of these values is less — i.e., an Accuracy Specification calling for ± 15% of the analytical reading, or ± 10% of the analyzer's full scale, whichever is less.

Precision

The Precision of any measurement will be the smallest quantity that the analytical instrument under consideration can indicate in its output reading. As an example, if the readout of an analyzer under consideration is in a digital format (i.e., 3.5 or 4.5 digits) showing two decimal places, then that analyzer's Precision would be 0.01 units. It is important to note that an analytical instrument's Precision is most assuredly not the same as its Sensitivity, although frequently these two parameters are mistaken and/or misunderstood to be identical.

Relevant Formulae and Relationships
Flow Rate and Flow Volume Calibrations

Flow rate calibrations are routinely performed using a combination of a volumetric standard in conjunction with a time standard. Simply, the time interval required for the output of some source of interest — i.e., a pump, etc. — to fill a precisely known volume is carefully measured and used then to determine the flow rate of the gas source.

Equation 2.1

$$\text{Flow Rate} = \frac{\text{Volume}}{\text{Time Interval}}$$

Where: **Flow Rate**   =   the volume of gas per unit time flowing in or out of some system, usually in units such as: liters/minute, cm^3/min, etc.;

Volume   =   the known standardized volume that has been filled in some known time interval, in units such as cm^3, or liters; and

| **Time Interval** | = | the actual measured time required for the gas source to output the standardized volume of gas, in some compatible unit such as minutes, etc. |

Equation 2.2

The principal purpose for making flow rate calibrations is to be able to calculate — with a high degree of certainty — the total volume of air that has been pumped, over a well-defined time interval, by a calibrated pump. These data are required for any determination of the average ambient concentration of any airborne material (gas, vapor, particulate, etc.) that might be trapped in any sort of impinger, filter cassette, etc. used in conjunction with the calibrated pump. Note that this relationship is simply a rearrangement of the previous equation.

$$\text{Total Volume} = [\text{Flow Rate}][\text{Time Interval}]$$

Where:	**Flow Rate**	=	the volume of gas per unit time flowing into or out of some system, as above, in units such as liters/minute;
	Total Volume	=	the calculated volume that has been pumped in some known time interval, in units such as liters; and
	Time Interval	=	the actual measured time interval during which the pump was in operation, in some compatible unit such as minutes, etc.

Gas Analyzer Calibrations and Calibration Checks

The process of calibrating, calibration checking, zeroing, span checking, etc. any gas analyzer is both a very necessary and relatively simple process. To accomplish this task, the individual involved must first develop a standard that contains a known and well-referenced concentration of the analyte of interest, and then use this standard to challenge the analyzer whose performance is to be documented.

Equations 2.3, 2.4, and 2.5

One of the most common methods for preparing a single concentration calibration standard that is to be used to test, calibrate, or span check a gas analyzer employs a chemically inert bag into which known volumes of a clean matrix gas (usually air or nitrogen) and a high purity analyte are introduced, so as to create a mixture of precisely known composition and concentration.

The sample preparation procedure always involves a minimum of two steps. First, a known volume of some matrix gas is introduced into a bag, inflating it to between 50 and 80% of its capacity. Next, a known volume of an analyte that is to serve as the standard is introduced into the bag. There are three very specific "categories" that apply to these single concentration calibration standards. Each will be described in detail in this section.

Equation 2.3

The first of the three equations is used when it is necessary to prepare and calculate the resultant concentration that arises from the introduction of *small volumes* of a *pure gas* into the matrix filled bag. This procedure is used whenever a low concentration level calibration standard — i.e., one in the ppm(vol) or ppb(vol) concentration range — is desired. Although the total volume in the chemically inert calibration bag will *always* consist of the volumes of both the matrix gas and the analyte, for calculation purposes, the analyte volume will be so extremely small that it can be ignored. This volume, which is typically measured in microliters, will be four to eight orders of magnitude smaller than the volume of the matrix gas, which, in contrast, will typically be measured in liters.

An important fundamental assumption in this overall process is that *all of the gas volumes* involved in every step of the preparation of the standard, and in completing the calculations that will identify the actual concentration in the standard, will have to have been normalized to some standardized set of conditions such as NTP or STP.

$$C = \frac{V_{analyte}}{V_{matrix}}$$

Where: **C**   =   the analyte concentration, in parts per million by volume;

           $V_{analyte}$ =   the volume of gaseous analyte that was introduced into the bag, measured in microliters; and

           V_{matrix} =   the precise volume of matrix gas introduced into the bag, measured in liters. As stated above, this matrix gas may be any pure gas (i.e., air, nitrogen, etc.) that, by definition, is completely free of impurities.

Equation 2.4

This second relationship is employed when the analyte is introduced as a gas into the bag or cylinder at sufficiently *large volumes* so as to produce a calibration standard, the concentration of which is most conveniently measured as a percent.

The very same important fundamental assumption that applied to Equation 2.3, also applies to this situation, namely, that *all of the gas volumes* involved in every step of the preparation of this standard, as well as in completing the calculations that will identify its actual concentration, will have to have been normalized to some standardized set of conditions such as NTP or STP.

$$C = 100\left[\frac{V_{analyte}}{1,000\left(V_{matrix}\right) + V_{analyte}}\right]$$

Where:

C = the analyte concentration, in percent by volume;

$V_{analyte}$ = the volume of gaseous analyte that was introduced into the bag, measured in milliliters; and

V_{matrix} = the precise volume of matrix gas introduced into the bag, measured in liters. As stated earlier, this matrix gas may be any pure gas (i.e., air, nitrogen, etc.) that, by definition, is completely free of impurities.

The final relationship is used whenever the calibration standard is to be prepared by the introduction of a known volume of a pure *liquid phase* chemical into the matrix filled bag. As was the case for standards produced by the introduction of a gaseous analyte, there are two concentration-related specific situations that will be covered — the first for low, and the second for high concentration level standards. In each of these cases, but particularly in the second or high concentration level case, care must be exercised to ensure that the prevailing conditions of temperature and pressure are sufficient to guarantee that *all* the liquid analyte will, in fact, vaporize so as to produce the desired concentration in the calibration standard. The relationship involved is the same for both cases.

Equation 2.5

$$C = \left[\frac{T_{ambient}V_{analyte}\rho_{analyte}}{16.036\left[P_{ambient}V_{matrix}\left(MW_{analyte}\right)\right] + T_{ambient}V_{analyte}\rho_{analyte}}\right]10^6$$

Where:

C = the analyte concentration, in parts per million by volume;

$T_{ambient}$ = the absolute ambient temperature, in K;

$V_{analyte}$ = the volume of pure liquid analyte introduced into the bag, measured in microliters, μl;

$\rho_{analyte}$ = the density of the pure liquid analyte, measured in grams/cm^3;

$P_{ambient}$ = the ambient barometric pressure, in mm Hg;

V_{matrix} = the precise volume of matrix gas introduced into the bag, measured in liters; and

$MW_{analyte}$ = the molecular weight of the analyte, measured in Atomic Mass Units (or more precisely, in grams mass per mole).

If the calibration standard to be generated by the introduction of a liquid into the bag must have its concentration in the percent range, then great care must be exercised to ensure that the prevailing conditions of temperature and pressure are sufficient to guarantee that <u>all</u> the liquid introduced will, in fact, evaporate so as to produce the desired analyte vapor concentration.

Note: In situations that involve the use of an inflatable bag, specific attention must be paid to the volume that the analyte — when completely vaporized from its liquid phase — will occupy. The injected volume of liquid will always be very small (i.e., it is measured in microliters); however, the analyte volume, when vaporized, will almost certainly be at least 2.5 to 3.0 orders of magnitude greater (i.e., 10 ml volume of acetone, introduced as a pure liquid, will vaporize to produce a gaseous volume of 3,325 ml = 3.33 liters at NTP — an obvious 330+ fold increase in volume). It is not at all uncommon, in the preparation of percentage concentration range standards by an individual who has overlooked this factor, to have a situation where the bag will burst when its capacity has been exceeded by the sum of the matrix gas and the vaporized analyte.

Chapter 3
Workplace Ambient Air

This chapter focuses on the wide variety of factors and aspects that characterize, or are characteristics of, ambient air. Particular emphasis is placed on those factors and/or conditions that must be mitigated in order to make the workplace safe. The principal components of any ambient matrix to be considered will be its temperature, pressure, and volume as well as the vapors, particulates, and/or aerosols that may be resident in it.

Relevant Definitions
Ambient Concentration Categories
Threshold Limit Values

The Threshold Limit Value (usually abbreviated, **TLV**) refers to an ambient airborne concentration of some substance of interest, and represents a condition under which it is believed that nearly *all* workers repeatedly may be exposed, day after day, without adverse effect. This concentration limit can be, and is, commonly expressed in one of three forms: as an 8-hour Time Weighted Average (TLV-TWA); as a Short Term Exposure Limit (TLV-STEL); and as a Ceiling Value (TLV-C). The overall ambient concentration concept designated by the term or phrase, "Threshold Limit Value," was introduced and promulgated by the American Conference of Government Industrial Hygienists (the **ACGIH**). Currently established TLVs are always under review, and individual listings are modified whenever relevant new information dictates that this be done.

Permissible Exposure Limits

The Permissible Exposure Limit (usually abbreviated, **PEL**) is an ambient airborne concentration of some substance of interest which the Occupational Safety and Health Administration (**OSHA**), which is a branch of the U.S. Department of Labor (**USDOL**), had adopted — largely from data furnished earlier by the ACGIH in the development of this organization's listing of Threshold Limit Values. The initial listing of these Permissible Exposure Limits was made in the Z-Tables of Title 29, Code of Federal Regulations (usually abbreviated **CFR**), Part 1910.1000, as published in the *Federal Register* on January 19, 1989. Currently published PELs are always under review; established values for specific materials are modified whenever additional information indicates that this should be done. In addition, PELs for previously unlisted materials are added as data on the effect of these materials are developed and become accepted. Like TLVs, PELs are commonly expressed in one of three forms: as an 8-hour Time Weighted Average (PEL-TWA); as a Short Term Exposure Limit (PEL-STEL); and as a Ceiling Value (PEL-C).

All PELs exist as enforceable statutes; whenever an employee or worker is exposed to a listed material at a combination of: (1) an ambient concentration, and/or (2) a duration of exposure that exceeds any of these specific standards, that individual's employer can be cited and fined by OSHA.

Recommended Exposure Limits

The Recommended Exposure Limit (usually abbreviated, **REL**) is still another ambient airborne concentration of some substance of interest which, in this case, the National Institute for Occupational Safety and Health (**NIOSH**) has researched and developed. Like its two previously listed counterparts, the TLV and the PEL, RELs are commonly expressed in one of the three standard forms: as an 8-hour Time Weighted Average (REL-TWA); as a Short Term Exposure Limit (REL-STEL); and as a Ceiling Value (REL-C). Currently established RELs are always under review, and individual listings are modified whenever relevant new NIOSH research dictates that this be done.

Maximum Concentration Values in the Workplace

The Maximum Arbeitsplatz Konzentration (usually abbreviated **MAK** — translated into English as the Maximum Concentration Value in the Workplace) is a TLV, PEL, and REL analog; and, as such, is also an ambient airborne concentration of some substance of interest which, in this case, the Deutsche Forschungsgemeinschaft (**DFG**), Commission for the Investigation of the Health Hazards of Chemical Compounds in the Work Area, as a branch of the Federal Republic of Germany's central government, has developed, adopted, and promulgated. Exactly like its three U. S. counterparts, MAKs are commonly expressed in one of the three standard forms: as an 8-hour Time Weighted Average (MAK-TWA), as a Short Term Exposure Limit (MAK-STEL), and as a Ceiling Value (MAK-C); and exactly like their U.S. counterparts, MAKs are always under review, and individual listings are modified whenever relevant new information dictates that this be done.

Time Weighted Averages

The Time Weighted Average (usually abbreviated as a "suffix", **-TWA**; thus: TLV-TWA, PEL-TWA, REL-TWA, and/or MAK-TWA) is the employee's average airborne exposure in *any* 8-hour work shift of *any* 40-hour work week to which nearly all workers may be repeatedly exposed, day after day, without suffering any adverse effects. It is a value that should never be exceeded.

Short Term Exposure Limits

The Short Term Exposure Limit (usually abbreviated as a "suffix", **-STEL**; thus: TLV-STEL, PEL-STEL, REL-STEL, and/or MAK-STEL) is the concentration to which workers can be continuously exposed for short periods of time without suffering from:

1. irritation;
2. chronic or irreversible tissue damage; or
3. narcosis of sufficient degree to increase the likelihood of accidental injury, impair self rescue, or materially reduce work efficiency — provided also that the corresponding **TWA** has not been exceeded.

STELs are usually 15-minute (except for those materials for which an alternative time limit has been specified) Time Weighted Average exposures which should *never* be exceeded during *any* of the specified time intervals during the work day, even if the corresponding TWA has not been exceeded. In the event any time limit other than 15 minutes is specified for some material or compound, the previous definition still holds, except as modified by the different time limit.

Ceiling Values

The **Ceiling Value** concentration (usually abbreviated as a "suffix", **-C**; thus: TLV-C, PEL-C, REL-C, and/or MAK-C) is a concentration that should *never* be exceeded, even instantaneously, at *any* time during the work day. In the event that instantaneous monitoring is not feasible, then the Ceiling Value can be assessed as a 15-minute Time Weighted Average exposure which should not be exceeded at any time during the work day, *except* when the subject vapor can cause immediate irritation with exceedingly short exposures.

Action Levels

The **Action Level** is an 8-hour Time Weighted Average concentration for which there is only a 5% risk of having more than 5% of the employee workdays involve an exposure at a level greater than the relevant TLV-TWA, PEL-TWA, REL-TWA, or MAK-TWA. This value is most frequently set at or near 50% of the relevant TLV-TWA, PEL-TWA, REL-TWA, or MAK-TWA concentration standard.

Excursion Limits

An **Excursion Limit** is a term that is frequently called into use in situations and for substances where no published STEL or Ceiling Value exists. It is a Short Term Exposure Limit or a Ceiling Value *without* any legal standing, as would be the case for OSHA published STELs or Ceiling Values. As such, the Excursion Limit is simply an "industry recognized" factor or guideline. For materials that have no published STEL, the STEL Excursion Limit is generally understood to be *three times* the published 8-hour Time Weighted Average standard, for no more than 30 minutes during any work day. For materials that have no published Ceiling Value, the Ceiling Value Excursion Limit is generally understood to be *five times* the established eight-hour Time Weighted Average standard, and is treated as a concentration that should *never* be exceeded at *any* time.

Designation of Immediately Dangerous to Life and/or Health

The Immediately Dangerous to Life and/or Health (usually abbreviated, **IDLH**) concentration is a concentration level of some substance of interest from which a worker could escape in 30 minutes or less without suffering any escape-impairing symptoms and/or irreversible health effects.

Breathing Zone

The **Breathing Zone** of an individual is a roughly hemispherical volume immediately forward of that person's shoulders and face, centered roughly on the Adam's Apple, and having a radius of 6 to 9 inches (15 to 23 cm).

Ambient Dose-Response/Concentration-Response Parameters

Median Lethal Dose

The Median Lethal Dose (usually abbreviated, **LD50**) is the toxicant dose at which 50% of a population of the same species will die within a specified period of time, under similar experimental conditions. This dosage number is usually expressed as milligrams of toxicant per kilogram of body weight (mg/kg).

Median Effective Dose

The Median Effective Dose (usually abbreviated, **ED50**) is the toxicant dose required to produce a specific nonlethal effect in 50% of a population of the same species, under similar experimental conditions. This dosage number is usually also expressed as milligrams of toxicant per kilogram of body weight (mg/kg).

Median Lethal Concentration

The Median Lethal Concentration (usually abbreviated, **LC50**) is the concentration of toxicant in air which will cause 50% of a population to die within a specified period of time. This factor is a concentration, not a dose, and is usually expressed either as milligrams of toxicant per cubic meter of air (mg/m^3), or as parts per million (ppm(vol)).

"Other" Dose-Response, Concentration-Response Parameters

Other common analogous Dose-Response or Concentration-Response Terms have been indicated below — in each case, the definition of each of these terms is directly analogous to the definition of the "Same Name" term on the previous page, the only difference being in the percentage figure involved:

Lethal Dose for **10%** of a population	(abbreviated, **LD10**)
Lethal Dose for **90%** of a population	(abbreviated, **LD90**)
Effective Dose for **10%** of a population	(abbreviated, **ED10**)
Effective Dose for **90%** of a population	(abbreviated, **ED90**)
Lethal Concentration for **10%** of a population	(abbreviated, **LC10**)
Lethal Concentration for **90%** of a population, etc.	(abbreviated, **LC90**)

Parameters Relating to Specific Chemicals or Substances

Upper and Lower Explosive Limits

The Upper and Lower Explosive Limits (abbreviated **UEL** and **LEL**) refer to the upper and lower vapor concentration boundaries, for some specific compound or material of interest, within which the vapor–air mixture will propagate a flame (i.e., explode) if ignited.

Explosive Range

The **Explosive Range** for any chemical or compound is the range of concentrations that exist between its Upper and Lower Explosive Limits. For example, the **LEL** and **UEL** for n-hexane are, respectively, 1.2 and 7.7%; thus this chemical's **Explosive Range** is 1.2 to 7.7% in air. When considering the **Explosive Range** for any specific chemical, it is important to note that for any concentration of that chemical that falls outside of this range, the vapor–air mixture possessing that concentration will not propagate a flame when a source of ignition is introduced. For chemical concentrations lower than the **LEL**, there is an insufficient amount of the combustible chemical to permit flame propagation, and for concentrations greater than the **UEL**, there is insufficient ambient oxygen to permit flame propagation.

Flash Point

The Flash Point of any compound is that temperature to which it must be heated before its vapors can be ignited by a free flame in the presence of air. It is a measure of the flammability of any material, and as such it is a reasonably good criterion for this characteristic. The lower the Flash Point, the more flammable a material is. This value is affected by the relative volatility _and_ the chemical composition of the material in question. Thus, ranked in the order of decreasing flammability, we would find the following to hold true (for a particular Flash Point designated as "**F**"):

$$F_{\text{pure hydrocarbons}} < F_{\text{oxygenated hydrocarbons}} < F_{\text{partially halogenated hydrocarbons}}, \text{ etc.}$$

As typical examples, $F_{\text{gasoline}} = -45°C$, $F_{\text{isopropanol}} = 12°C$, and $F_{\text{lubricating oil}} = 232°C$, etc. Certain materials (i.e., carbon tetrachloride, CCl_4) do not have a Flash Point, since there is _no_ temperature at which their vapors can be ignited.

Ambient Measurements of Concentration
"Volume-Based" Concentrations

Ambient concentrations are most frequently expressed in, and understood on, a basis that is understood to be:

a unit volume-per-multiple volumes
or, more precisely
a unit molecule-per-multiple molecules basis.

Among the most commonly used volume-based concentration units are each of the following:

Concentration Unit	*Form in which these concentration units commonly are expressed*	*Volume(or Molecular) Definition*
parts per million	**ppm(vol)** *or* simply **ppm**	Volumes (or molecules) of a specific material of interest *per* million volumes (or *per* million molecules) present in the total matrix being considered.
parts per billion	**ppb(vol)** *or* simply **ppb**	Volumes (or molecules) of a specific material of interest *per* billion volumes (or *per* billion molecules) present in the total matrix being considered.
percent	%	Volumes (or molecules) of a specific material of interest per hundred volumes (or per hundred molecules) present in the total matrix being considered.

While there are other volume-based concentration units (i.e., parts per hundred thousand, etc.) they are far less common than any of the foregoing three.

Two very important factors and/or relationships *must be understood* about any volume-based concentration unit. These are:

1. The volume-based concentration unit can only be used to express the ambient concentration of a gas or vapor; it can never be used to express the ambient concentration of any form of particulate, dust, or aerosol.

2. Virtually all ambient gas analyzers provide concentration outputs in the form of "parts per million by volume" (ppm(vol)). With very rare exceptions, the calibrated output of these analyzers will always have been referenced to the readout that would have been obtained for measurements made at sea level. If a gas analyzer is used at an altitude that is significantly different than sea level, its readout will be incorrect in proportion to the difference in the actual barometric pressures (1) at

the location where the analyzer has been used, and (2) at sea level — see **Appendix A**, which goes into more detail on the atmosphere and its effects on a wide variety of measurements.

"Mass-Based" Concentrations

Ambient concentrations are less frequently expressed in, and understood on, the basis of the mass of material of interest-per-unit volume in the ambient matrix being considered. Virtually without exception, the two principal mass-based units of concentration, as applied to the ambient air, are:

milligrams of the material of interest-per-cubic meter of matrix
mg/m³ and

micrograms of the material of interest-per-liter of matrix
µg/l

For any situation that is being evaluated, the magnitude of the numeric value that would be obtained for either of these two mass-based ambient concentrations would be the same — i.e., if it is known that a workplace has an ambient silica dust concentration of 0.13 mg/m³, it would turn out that the process of converting this concentration to the form of micrograms-per-liter would produce the numerically identical value, namely, 0.13 µg/l.

From the perspective of the potential adverse impacts that might be caused by some material that exists as a component of the ambient air, the mass-based concentration of this material is a somewhat more "absolute" parameter than its volume-based counterpart. To determine the ambient concentration of any particulate, dust, or aerosol in the ambient air, the mass-based concentration parameter is the *only* form that is used.

Any analyzer that provides ambient mass-based concentration outputs (for gases, vapors, and/or particulates) will provide correct values irrespective of the altitude where the measurements are made.

Components and Measurement Parameters of the Ambient Air

Gas

A **Gas** is a substance that is in the gaseous state at **NTP**.

Vapor

A **Vapor** is the gaseous state of any material that would, under **NTP**, exist principally as a solid or a liquid.

Aerodynamic Diameter

The **Aerodynamic Diameter** is the diameter of a unit density sphere (i.e., density = 1.00 gms/ cm^3) that would have the same settling velocity as the particle or aerosol in question.

Aerosol

An **Aerosol** is a suspension of liquid or solid particles in the air. Actual physical particle diameters usually fall in the range:

$$0.01\mu \leq \text{(particle diameter)} \leq 100\mu.$$

Dust

Dust is any particulate material, usually generated by a mechanical process, such as crushing, grinding, etc. Typical dust particles have aerodynamic diameters in the range:

$$0.5 \ \mu m \leq \text{(aerodynamic diameter)} \leq 50 \ \mu m.$$

Mist

A **Mist** is an aerosol suspension of liquid particles in the air, usually formed either by condensation directly from the vapor phase or by some mechanical process. Typical mist droplets have aerodynamic diameters in the range:

$$40\mu m \leq \text{(aerodynamic diameter)} \leq 400 \ \mu m.$$

Smoke

Smoke is an aerosol suspension, usually of solid particulates, and usually formed by either the combustion of organic materials or the sublimation of some material. Typical smoke particulates have aerodynamic diameters in the range:

$$0.01\mu m \leq \text{(aerodynamic diameter)} \leq 0.5 \ \mu m.$$

Fume

A **Fume** is an aerosol made up of solid particulates formed by condensation directly from the vapor state. Typical fume particles have aerodynamic diameters in the range:

$$0.001\mu m \leq \text{(aerodynamic diameter)} \leq 0.2 \ \mu m.$$

Aspect Ratio

The **Aspect Ratio** for any particle is the ratio of its longest or greatest dimension to its shortest or smallest dimension.

Fiber

A **Fiber** is any particle having an aspect ratio greater than 3.

Relevant Formulae and Relationships
Calculations Involving Time Weighted Averages (TWAs)
Equation 3.1

Equation 3.1 is used to obtain a **Time Weighted Average** for any sort of time related exposure or measurement; it applies to any Time Weighted Average determination and can be utilized for any total base time period or time interval. Typically this calculation will be made in order to develop a TWA exposure value for use in comparison against any of the currently published exposure limit standards (i.e., the TLV, the PEL, the REL, and/or the MAK). As an example, any Short Term Exposure Limit exposure determination will use this Equation. The denominator for any such calculation will be determined by the specified STEL Time Interval that applies to the material of interest in the determination (i.e., if the material being evaluated — from the perspective of referencing against its established PEL-STEL — were to have been isopropanol, then the STEL Time Interval would have been 15 minutes):

$$\text{TWA} = \frac{\sum_{i=1}^{n} T_i C_i}{\sum_{i=1}^{n} T_i} = \frac{T_1 C_1 + T_2 C_2 + \ldots + T_n C_n}{T_1 + T_2 + \ldots + T_n}$$

Where: **TWA** = The Time Weighted Average Concentration that existed during the Time Intervals given by the sum of the individual T_is used in this calculation, i.e., the

$$\sum_{i=1}^{n} T_i.$$

T_i = the ith Time Interval from the overall time period, which for a TWA (in contrast to a STEL) would total up to a full 8 hours, i.e.,

$$\sum_{i=1}^{n} T_i = 8 \text{ hours.}$$

C_i = the ith Concentration Value of the single component of interest — i.e., the ambient gas or particulate

concentration — that existed during
the specific ith Time Interval.

Equation 3.2

Equation 3.2 is used to determine the **Effective Percent Exposure Level** for any of the established parameters (i.e., the TLV, the PEL, the REL, or the MAK), resulting from the combined effects of *all* the potentially irritating, toxic, or hazardous components, whether gas or particulate, in any ambient air system that is being evaluated. To repeat, it is effective for any ambient air matrix, whether there is only a single volatile or aerosolized component in it, or many such components.

In applying Equation 3.2, the individual performing the calculation must know the specific value of the Standard being applied (i.e., for the case shown, the REL) that has been established for each of the various components that are contained in the ambient air that is being evaluated. If, for example, the evaluation were to involve an 8-hour Recommended Exposure Limit – Time Weighted Average determination for an air mass containing three different refrigerants: R-12, R-22, and R-112, then the REL-TWAs for these three compounds would have to be known. For reference in this example, these REL-TWA Values are as follows: for R-12, the **REL-TWA$_{R-12}$** = 1,000 ppm(vol); for R-22, the **REL-TWA$_{R-22}$** = 1,000 ppm(vol); and finally for R-112, the **REL-TWA$_{R-112}$** = 500 ppm(vol). The effective Time Weighted Average Concentration Values (i.e., the TWA$_i$s) that are shown in the numerator of Equation 3.2 would be determined by applying Equation 3.1 to the available information or data.

$$\% \text{ REL} = 100 \left[\sum_{i=1}^{n} \frac{\text{TWA}_i}{\text{REL}_i} \right] = 100 \left[\frac{\text{TWA}_1}{\text{REL}_1} + \frac{\text{TWA}_2}{\text{REL}_2} + \cdots + \frac{\text{TWA}_n}{\text{REL}_n} \right]$$

Where: **% REL** = the **Effective Percent Exposure Level** from the perspective of the Recommended Exposure Limit Standards that was achieved for the mixture being evaluated, expressed always as a percentage.

TWA$_i$ = the Time Weighted Average Concentration of the ith component in the mixture being evaluated.

REL$_i$ = the listed **Recommended Exposure Limit** (or, in this case, the **REL-TWA**) of the ith component.

Note: If the **% REL** is equal to or less than 100%, then it can be inferred that the

Effective REL for the mixture has not been exceeded; if this % REL is greater than 100%, then the inference is that the Effective REL for the mixture has been exceeded.

Calculations Involving Exposure Limits
Equation 3.3

Equation 3.3 is used to determine the **Effective Exposure Limit** that exists for any equilibrium vapor phase that is in contact with any well-defined liquid mixture containing two or more different volatile components. It can be applied to any of the established parameters (i.e., the TLV, the PEL, the REL, or the MAK); the value of the individual Exposure Limit Standard for each of the components in the liquid mixture must be known. If this **Effective Exposure Limit** is determined *either* by using different Exposure Limit Standards (i.e., TLV-TWAs coupled with PEL-TWAs), OR by using the same Exposure Limit Standard, with different time interval bases (i.e., the REL-STEL$_{hydrazine}$ which must be evaluated over a 120-minute period, and the REL-STEL$_{1,2-dioxane}$ which must be evaluated over a 30-minute period), then the resultant calculated **Effective Exposure Limit** would be considered to be *neither* valid *nor* accurate.

Equation 3.3 assumes that the overall composition of the vapor phase existing above a volatile liquid mixture will be *identical* to the mass composition of the liquid mixture, and although this would rarely — if ever — be true, the equation is considered to be a useful tool for determining "Order-of-Magnitude-Approximations" for the concentrations of the components in the vapor phase. This Equation was proposed for use by the ACGIH; therefore, it is used most commonly for the determination of this organization's Exposure Limit Standard — namely, the TLV; it can, however, be applied to any of the commonly used Exposure Limit Standards.

If the user wishes a more precise treatment of this situation, then what is required would be the sequential application of several of the more basic *laws and relationships*, each of which has been documented in Chapter 1, which covered the basic laws of physics and the basic gas laws. For reference, these would include Dalton's Law of Partial Pressures (Equations 1.15 and 1.16, and Raoult's Law (Equations 1.17 and 1.18).

$$TLV_{effective} = \left[\frac{1}{\sum\limits_{i=1}^{n}\frac{f_i}{TLV_i}}\right] = \left[\frac{1}{\frac{f_1}{TLV_1} + \frac{f_2}{TLV_2} + \ldots + \frac{f_n}{TLV_n}}\right]$$

Where: $TLV_{effective}$ = the **Effective Exposure Limit** — in this case a Threshold Limit Value.

This parameter can be evaluated for *any* of the established Exposure Limits — i.e., for a PEL, REL, or MAK, etc. Remember, however, that this relationship applies *only* to the vapor phase existing above a mixture of volatile liquids. The user must understand that the result of this calculation is an "Order-of-Magnitude-Approximation" *only*. Also, in this case the calculated **TLV**$_{effective}$ always will be expressed in mass concentration units, mg/m^3, NEVER the more common volumetric unit, ppm(vol).

f$_i$ = the Weight Fraction of the ith component in the mixture of volatile liquid components being considered, i.e., the

$$\left[\frac{\text{mass of the component of interest}}{\text{total mass of the liquid mixture}} \right];$$

TLV$_i$ = the Threshold Limit Value (TLV) — this could be any established Exposure Limit — of the ith component, must always be expressed in mass-based units of concentration, mg/m^3, NEVER the more common volume-based unit, ppm(vol).

Calculations Involving the Conversion of Concentration Units

Equations 3.4, 3.5, and 3.6

The following three equations, 3.4, 3.5, and 3.6, are used to convert sets of **Mass-Based Concentration** units to their **Volume-Based Concentration** equivalents (e.g., converting from concentrations expressed in units such as mg/m^3 to those in one of the common volumetric-based sets of units, such as ppm(vol)). The first equation, 3.4, would be used to affect this conversion under conditions of Normal Temperature and Pressure (NTP); the second, 3.5, would be used for conversions at Standard Temperature and Pressure (STP) conditions, while the third, 3.6, applies to any mass-to-volume-based concentration conversion, regardless of the conditions of ambient temperature and pressure.

Equation 3.4

$$C_{vol} = \frac{24.45}{MW_i}\left[C_{mass}\right] \ @ \ \mathbf{NTP}$$

Where: C_{vol} =   the Volume-Based Concentration of the component of interest, namely, the ith component, measured in ppm(vol);

           C_{mass} =   the Mass-Based Concentration of the same component, namely the ith component, in mg/m^3; and

           MW_i =   the Molecular Weight of the same ith component.

Equation 3.5

$$C_{vol} = \frac{22.41}{MW_i}\left[C_{mass}\right] \ @ \ \mathbf{STP}$$

Where: C_{vol} =   the Volume-Based Concentration of the component of interest, namely, the ith component, measured in ppm(vol);

           C_{mass} =   the Mass-Based Concentration of the same component, namely the ith component, in mg/m^3; and

           MW_i =   the Molecular Weight of the same ith component.

Equation 3.6

$$C_{vol} = \frac{R\,T}{P\left[MW_i\right]}\left[C_{mass}\right]$$

under *any* conditions of ambient Temperature and Pressure

Where: C_{vol} =   the Volume-Based Concentration as defined immediately above for Equations 3.4 and 3.5, in ppm(vol);

           C_{mass} =   the Mass-Based Concentration also as defined immediately above for Equations 34 and 3.5, in mg/m^3;

MW_i = the Molecular Weight of the ith component, again as defined above;

T = the ambient Absolute Temperature, in some suitable units of absolute temperature, most probably, K;

P = the ambient Barometric Pressure, in some suitable units of pressure, such as mm Hg; and

R = the Universal Gas Constant in units consistent with those used for the prevailing ambient barometric pressure and temperature conditions — for example, any of the following values of this Universal Gas Constant could be used.

$$= \quad 0.0821 \ \frac{(\text{liter})(\text{atmospheres})}{(\text{K})(\text{mole})};$$

$$= \quad 62.36 \ \frac{(\text{liter})(\text{mm Hg})}{(\text{K})(\text{mole})};$$

$$= \quad 1.631 \ \frac{(\text{feet})^3(\text{millibars})}{(°\text{R})(\text{mole})}.$$

Equations 3.7, 3.8, and 3.9

The following three Equations, 3.7, 3.8, and 3.9, are used to convert sets of Volume-Based Concentration units to their Mass-Based Concentration equivalents (e.g., converting from concentrations expressed in ppm (vol) to those expressed in mg/m³). The first equation, 3.7, would be used to affect this conversion under Normal Temperature and Pressure (NTP) conditions; the second, 3.8, would be used under Standard Temperature and Pressure (STP) conditions; while the third, 3.9, applies in general to any mass-to-volume-based concentration conversion, regardless of the conditions of ambient temperature and pressure.

Equation 3.7

$$C_{mass} = \frac{MW_1}{24.45}[C_{vol}] \ @ \ NTP$$

Where: C_{vol} = the Volume-Based Concentration of the component of interest, namely, the **ith** component, measured in ppm(vol);

C_{mass} = the Mass-Based Concentration of the same component, namely the **ith** component, in mg/m^3; and

MW_i = the Molecular Weight of the same **ith** component.

Equation 3.8

$$C_{mass} = \frac{MW_1}{22.41}\left[C_{vol}\right] \text{ @ } \textbf{STP}$$

Where: C_{vol} = the Volume-Based Concentration of the component of interest, namely, the **ith** component, measured in ppm(vol);

C_{mass} = the Mass-Based Concentration of the same component, namely the **ith** component, in mg/m^3; and

MW_i = the Molecular Weight of the same **ith** component.

Equation 3.9

$$C_{mass} = \frac{P\left[MW_1\right]}{R\,T}\left[C_{vol}\right] \text{ under } any \text{ conditions of ambient}$$
temperature and pressure

Where: C_{vol} = the Volume-Based Concentration as defined for Equations **7** and **8**, in ppm(vol);

C_{mass} = the Mass-Based Concentration also as defined immediately above for Equations 3.7 and 3.8, in mg/m^3;

MW_i = the Molecular Weight of the **ith** component, again as defined for Equations 3.7 and 3,8. in ppm(vol;

T = the ambient Absolute Temperature, in some suitable units of absolute temperature, most probably, K;

P	=	the ambient Barometric Pressure, in some suitable units of pressure, such as mm Hg; and
R	=	the Universal Gas Constant in units consistent with those used for the prevailing ambient barometric pressure and temperature conditions — for example, any of the following values of this Universal Gas Constant could be used.

$$= \quad 0.0821 \; \frac{(\text{liter})(\text{atmospheres})}{(\text{K})(\text{mole})};$$

$$= \quad 62.36 \; \frac{(\text{liter})(\text{mm Hg})}{(\text{K})(\text{mole})};$$

$$= \quad 1.631 \; \frac{(\text{feet})^3(\text{millibars})}{(°\text{R})(\text{mole})}.$$

Calculations Involving TLV Exposure Limits for Free Silica Dusts in Ambient Air
Equations 3.10, 3.11, and 3.12

The following three equations, 3.10, 3.11, and 3.12, are used now only very infrequently, and only when one must develop a rough approximation of the respective required TLVs. These Equations were developed and promulgated by the ACGIH, and are used, therefore, only to determine Threshold Limit Value Exposure Limits. In order, these three Equations are as follows: 3.10 — For Respirable Quartz Dusts; 3.11 — For Total Dusts; and 3.12 — For Mixtures of the Three Most Common Types of Silica Dusts. These three equations are shown here.

Equation 3.10

$$\boxed{\text{TLV}_{\text{quartz}} = \left[\frac{10\text{mg}/\text{m}^3}{\%\text{RQ} + 2} \right]}$$

Where: **TLV**_{quartz}	=	the calculated 8-hour TWA Threshold Limit Value for respirable quartz dust, in mg/m³; and
%RQ	=	the Mass or Weight Fraction of Respirable Quartz Dusts in the sample being evaluated, expressed as a percentage.

Equation 3.11

$$TLV_{dust} = \left[\frac{30mg / m^3}{\%Q + 2} \right]$$

Where: **TLV**$_{dust}$ =   the calculated 8-hour TWA Threshold Limit Value for total dusts, in mg/m^3; and

  %Q   =   the Mass or Weight Fraction of Quartz Dusts in the sample being evaluated, expressed as a percentage.

Equation 3.12

$$TLV_{mix} = \left[\frac{10mg / m^3}{\%Q + 2(\%C) + 2(\%T) + 2} \right]$$

Where: **TLV**$_{mix}$ =   the calculated 8-hour TWA Threshold Limit Value for the complex mixture of silica dusts, in mg/m^3;

  %Q   =   the Mass or Weight Fraction of Quartz Dusts in the sample being evaluated, expressed as a percentage;

  %C   =   the Mass or Weight Fraction of Cristobalite Dusts in the sample being evaluated, also expressed as a percentage; and

  %T   =   the Mass or Weight Fraction of Tridymite Dusts in the sample being evaluated, this one, also, expressed as a percentage.

Chapter 4
Ventilation

This chapter will discuss the parameters, measurements, mathematical relationships, and the various pieces of hardware, equipment, etc., that constitute the general area of ventilation.

Relevant Definitions
Ventilation Equipment and Basic Parameters

Ventilation is an operation whereby air is moved to or from various locations within a facility by and/or through a variety of different pieces of equipment or hardware. Selecting from the wide variety of available ventilation equipment will always be a function of the task that is to be accomplished. Tasks such as providing fresh air to a office, removing vapors or fumes produced by some manufacturing process, etc. are all recognizable objectives that can be readily achieved by the proper choice and operation of appropriate ventilation equipment. The various items of hardware, and the factors and parameters that characterize each, will be defined and described below.

Hood

A **Hood** is a shaped inlet designed and positioned so as to capture air that contains some sort of contaminant, and then conduct this contaminated air into the ducts of the exhaust system. Hoods may be plain or flanged depending upon a variety of design factors.

Flange

A **Flange** is generally a flat piece of metal or rigid plastic that surrounds the shaped inlet of a hood and serves the purpose of minimizing the drawing of air from nearby zones where no contaminant is known or thought to exist.

Capture Velocity

The **Capture Velocity** is the air velocity at any point in front of a hood, or hood opening, that is necessary to overcome the opposing air currents and other factors, and thereby "capture" the contaminated air at that point by causing it to flow or be drawn into the hood.

Slot Velocity

The **Slot Velocity** is the actual linear velocity of air as it enters a hood through its slot opening.

Face Velocity

The **Face Velocity** is the linear velocity of air, averaged over the entire opening, as it enters a hood.

Coefficient of Entry

The **Coefficient of Entry** is the actual rate of flow produced by a given hood static pressure, when compared to the theoretical flow that would result if the static pressure in the hood could be converted, with 100% efficiency, into the velocity pressure of the fluid being drawn into the hood. The **Coefficient of Entry** is the ratio of the actual flow rate to the theoretical maximum flow rate.

Entry Loss

Entry Loss is the loss or decrease in pressure experienced by air as it flows into or enters a duct or hood. **Entry Losses** are usually expressed in "inches of water-gauge."

Duct

A **Duct** is a hollow, cylindrical (circular, rectangular, etc.) connected section of tubing that serves as the main conduit for the flow of air in any ventilation system.

Blast Gate or Damper

A **Blast Gate** or **Damper** is a sliding, swinging, or pivoting "door" that can shut off or decrease the cross sectional area of a duct, thereby limiting the flow of air through it. The principal purpose of a **Blast Gate** or **Damper** is, obviously, to permit the easy adjustment of the volume of air that is flowing through the duct. Adjustments of this type are usually required in order to improve or balance the general flow of ventilation air to all locations.

Plenum

A **Plenum** is a chamber within an overall ventilation system that serves the purpose of equalizing potentially different pressures that might have arisen from various different streams of air as they merge or separate in an overall system of ducting.

Flow Rate or Volumetric Flow Rate

Flow Rate or **Volumetric Flow Rate** is a measure of the volume of a gas (usually air) that is moving through a fan or a duct. Its most commonly used dimensions are "volume per unit time" in units such as cubic feet per minute (cfm), etc.

Velocity or Duct Velocity

Velocity or **Duct Velocity** is a measure of the speed with which a gas (again, usually air) is moving through a duct. This velocity is always considered to be in a direction that is perpendicular to the cross section of the duct through which the gas is flowing. Its most commonly used dimensions are "distance per unit time" in units such as feet per minute (fpm).

Air Handler or Blower or Fan

An **Air Handler, Blower**, or **Fan** is the basic or prime motive force that is integral to every ventilation system. The movement of air in any such system will always have been caused by such a prime mover.

Air Horsepower

The **Air Horsepower** is the theoretical horsepower required to operate a fan, blower, or air handler on the assumption that this item of hardware can be operated without any internal losses (i.e., the horsepower that would be required to operate a 100% efficient, totally friction free unit).

Brake Horsepower

The **Brake Horsepower** is the horsepower actually required to operate a fan. **Brake Horsepower** involves *all* the internal losses in the fan and can only be measured by conducting an actual test on the fan.

Pressure Terms

To understand the area of ventilation, one must have an in-depth knowledge of the various types of pressure measurements that characterize this general subject. Pressure terms will be defined and discussed in detail.

Static Pressure

Static Pressure is the pressure exerted in all directions by a fluid at rest. For any fluid in motion, this parameter is measured in a direction normal to the direction of fluid flow. **Static Pressure** is usually expressed in "inches of water-gauge" when dealing with air flowing in a duct; this parameter can be either negative or positive.

Velocity Pressure

Velocity Pressure is the kinetic pressure that is necessary in order to cause a fluid to flow at a particular velocity; it is measured in the direction of fluid flow. **Velocity Pressure** is usually expressed in "inches of water-gauge" when dealing with air flowing in a duct; this parameter is always positive. It is measured in a direction *parallel* to the direction of fluid flow, looking "upstream," or directly into the flowing fluid.

Total Pressure

Total Pressure is the algebraic sum of the Static and the Velocity Pressures. When dealing with air flowing in a duct, **Total Pressure**, like the other ventilation related pressure terms, will usually be expressed in "inches of water-gauge."

Manometer

A **Manometer** is an instrument that is used for measuring the pressure in a stationary or flowing fluid or gas. A **Manometer** very often is a U-tube filled with water, light oil, or mercury. This implement usually is constructed so that the observed displacement of liquid in it will indicate the pressure being exerted on it by the fluid being monitored. In a Static Pressure measuring situation, the *plane* of the opening of the Manometer tube in the fluid being monitored will be *parallel* to the direction of flow of that fluid (i.e., the Manometer tube, itself, will be situated *perpendicular* to the flow of that fluid). In a Velocity Pressure measuring situation, the *plane* of the opening of the Manometer tube in the fluid being monitored will be *perpendicular* to the direction of flow of that fluid, and will be directed so that the opening at the end of the tube faces directly into that flow (i.e., the Manometer tube, itself, will be situated *parallel* to the flow of that fluid, with the tube opening facing "upstream").

Vapor Pressure

Vapor Pressure is the static pressure exerted by a vapor. If the vapor phase of some material of interest is in equilibrium with its liquid phase, then the **Vapor Pressure** that exists is referred to as the Saturated Vapor Pressure for that material. The saturated vapor pressure for any material is solely dependent on its temperature. Frequently vapor pressure and saturated vapor pressure are used synonymously.

Humidity Factors

Humidity, or the presence of water vapor in the air in a structure or flowing in a duct, is always a source of concern to individuals who must work with ventilation systems. Several measures of Humidity will be discussed in this section.

Absolute Humidity

Absolute Humidity is the weight of the vaporous water in the air measured per unit volume of that air, usually expressed in lbs/ft^3 or $grams/cm^3$.

Relative Humidity

Relative Humidity is the ratio of the actual or measured vapor pressure of water in an air mass to the saturated vapor pressure of pure water at the same temperature as that of the air mass. **Relative Humidity** also can be thought of as the percentage that the actual measured absolute concentration of water in the air under *any* ambient conditions is to the saturation concentration of water in air under *the same* ambient conditions.

Dew Point

The **Dew Point** of an air mass is also a measure of relative humidity. It is the temperature at which the water vapor present in an air mass would start to condense — or in the context of the term, **Dew Point**, it is the temperature at which that water vapor would turn to "dew". It can be determined experimentally simply by cooling a mass of air until "dew" or condensate forms; the temperature at which this event occurs is defined to be the **Dew Point** of that air mass.

Ventilation Standards
Ventilation Standard Air

Standard Air, as Ventilation Engineers view it, differs somewhat from air at STP or NTP (as defined earlier). **Standard Air** is air possessing the following characteristics:

Temperature:	70°F
	21.1°C
	529.67°R
	294.26°K
Barometric Pressure:	1.00 atmosphere
	760 mm Hg
	29.92 in. of Hg
	14.70 psia
	0.00 psig
	1,013.25 millibars
	760 Torr
Humidity Level:	Relative Humidity = 0.0%,
	Absolute Humidity = 0.0 grams/cm^3
	Water Vapor Concentration = 0.0 ppm(vol) = 0.0 mg/m^3
Specific Gravity:	Density = 0.075 lbs/ft^3
Heat Capacity:	Specific Heat = 0.24 BTU/lb/°F.

Relevant Formulae and Relationships
Calculations Involving Gases Moving in Ducts

Most of the important calculations in the area of ventilation involve some gas, usually air, that is being conducted through ducting in order to achieve one or more specific purposes. The relationships that follow permit the engineer, whose responsibility it is to operate a ventilation system so as to accomplish the stated goals or tasks, to determine exactly how his system can be operated in order to achieve the designed purposes or goals.

Equation 4.1

Equation 4.1 is probably the most basic relationship in the general area of ventilation. It relates the **Volumetric Flow Rate** in a duct to the **Cross Sectional Area** of that duct, and to the **Velocity** (the **Duct Velocity**) of the gas flowing in it.

$$Q = AV$$

Where: Q   =   the **Volumetric Flow Rate** in the duct, measured in cubic feet/minute = ft³/min (cfm);

            A   =   the **Cross Sectional Area** of the duct under consideration, in square feet (ft²); and

            V   =   the **Velocity**, or **Duct Velocity**, of the gases moving in the duct, in feet/minute = ft/min (fpm).

Equation 4.2

Equation 4.2 is the first of two basic relationships involving the **Velocity** of the gases that are flowing in a duct. This expression relates this parameter to the **Velocity Pressure** and the **Density** of the gases that are flowing in the duct.

$$V = 1,096 \sqrt{\frac{VP}{\rho}}$$

Where: V   =   the **Velocity** of the gases moving in the duct, or the **Duct Velocity**, measured in feet per minute (fpm);

            VP =   the **Velocity Pressure** of the gases moving in the duct, measured in inches of water; and

            ρ   =   the **Density** of the gases that are flowing in the duct, measured in lbs per cubic foot (lbs/ft³) — for reference, the density of Ventilation Standard Air = 0.075 lbs/ft³.

Equation 4.3

Equation 4.3 is the second of the two basic relationships involving the **Duct Velocity** of the gases that are flowing in a duct. This expression, which simply relates this parameter to the **Velocity Pressure** of the gases that are flowing in the duct, is shown below.

$$V = 4,005\sqrt{VP}$$

Where: V   =   the **Velocity** of the gases moving in the duct, or the **Duct Velocity**, measured in feet per minute (fpm);

            VP =   the **Velocity Pressure**, and is as defined for Equation 4.2.

The following three equations, namely, 4.4, 4.5, and 4.6, are presented as a group — they are known as the *General Rules of Ventilation* — because: (1) they constitute three relationships that will always apply to gases — usually air — that happen to be flowing in a duct, and (2) they are closely interrelated.

Equation 4.4

$$TP_1 = TP_2 + \text{losses}$$

Where:			
TP_1	=	the **Total Pressure** at Point 1 in the duct, measured in inches of water;
TP_2	=	the **Total Pressure** at Point 2 — which is downstream of Point 1 — in the duct, also measured in inches of water; and
losses	=	Pressure Losses from various duct internal factors (i.e., friction) that affect the flow of gases passing through the duct — these total losses are almost always measured in inches of water. In general, the factors that can be used to quantify these losses will be provided by the duct's manufacturer, usually in the form of a "Specific Pressure Loss or Pressure Drop per Unit Length of Duct." These design factors will be expressed in units of "inches of water loss per lineal foot of duct traversed."

Equation 4.5

$$TP = SP + VP$$

Where:			
TP	=	the **Total Pressure** in the duct, and is as defined for Equation 4.4, where it was listed either as TP_1 or TP_2, and measured in inches of water;
SP	=	the **Static Pressure** in the duct, usually measured in inches of water; and
VP	=	the **Velocity Pressure** in the duct, and is exactly as defined for Equation 4.2.

Equation 4.6

$$SP_1 + VP_1 = SP_2 + VP_2 + \text{losses}$$

Where: **SP$_i$**   =   the **Static Pressure** at the i:h point in the duct, and is as defined for Equation 4.5, where it is listed simply as **SP**;

VP$_i$ =   the **Velocity Pressure** at the ith point in the duct, and is as defined for Equation 4.5, where it is listed simply as **VP**; and

losses   =   the **Pressure Losses** and are exactly as defined for Equation 4.4.

Calculations Involving Hoods

The following two equations, 4.7 and 4.8, are the relationships that provide the **Capture Velocity** — i.e., the hood opening centerline velocity — for either a square or a round hood opening having a known cross sectional area — for hoods either without Flanges (Equation 4.7), or those *with* Flanges (Equation 4.8) — with the evaluation of this velocity being made at some specified distance directly in front of the plane of the hood opening being considered and evaluated.

Equation 4.7 (Hoods *without* Flanges)

$$V = \frac{Q}{10x^2 + A}$$

Where: **V**   =   the **Capture Velocity** — i.e., the centerline velocity of the air entering the hood under consideration, at a point "**x**" feet directly in front of the face of the hood. This **Capture Velocity** is usually measured in feet per minute (fpm);

Q   =   the **Volumetric Flow Rate** of the hood, measured in cubic feet per minute (cfm);

A   =   the **Cross Sectional Area** of the hood opening, measured in square feet (ft^2); and

x   =   the **Distance** from the plane of the hood opening to the point directly in front of it where the **Capture Ve-**

locity is to be determined, measured
in feet (ft).

Equation 4.8 (Hoods *with* Flanges)

$$V = \frac{Q}{0.75\left[10x^2 + A\right]} = \frac{4Q}{3\left[10x^2 + A\right]}$$

Where: V, Q, A, and x are all exactly as defined for Equation 4.7.

Equation 4.9

Equation 4.9 is usually referred to as the *Simple Hood Formula*. It is widely used to determine the **Hood Static Pressure**, which is one of the most important operating factors applying to the operation of any hood.

$$SP_h = VP_d + h_e$$

Where: SP_h = the **Hood Static Pressure**, measured usually in inches of water;

VP_d = the **Velocity Pressure** in the hood duct, also measured in inches of water; and

h_e = the **Hood Entry Losses**, also measured in inches of water.

Equation 4.10

Equation 4.10 is used to calculate the **Coefficient of Entry** for any hood. This dimensionless parameter serves, functionally at least, as a Hood Efficiency Rating.

$$C_e = \sqrt{\frac{VP_d}{SP_h}} = \frac{\text{Actual Flow}}{\text{Theoretical Flow}}$$

Where: C_e = the **Coefficient of Entry** for the hood under investigation, this is a dimensionless parameter;

VP_d = the **Velocity Pressure** in the hood duct, and is as defined above on this page for Equation 4.9; and

SP_h = the **Hood Static Pressure**, and also is as defined for Equation 4.9.

Equation 4.11

The following two forms of Equation 4.11 are known, collectively, as the *Hood Throat-Suction Equations*. They, too, are widely used to determine the volumetric flow rates of hoods.

$$Q = 4,005A\sqrt{VP_d} \quad \text{— Form 1}$$

$$Q = 4,005AC_e\sqrt{SP_h} \quad \text{— Form 2}$$

Where:

Q = the **Volumetric Flow Rate** and is as defined for equations 4.7 and 4.8;

A = the **Cross Sectional Area** of the hood opening, and is also as defined for Equations 4.7 and 4.8;

VP_d = the **Velocity Pressure** in the hood duct, and is as defined for Equation 4.9;

C_e = is the **Coefficient of Entry**, and is also as defined for Equation 4.10; and

SP_h = is the **Hood Static Pressure**, and is also as defined for Equation 4.10.

Equation 4.12

This Equation 4.12 is known as the *Hood Entry Loss Equation*, and simply uses most of the previously defined parameters to develop a different but very useful relationship.

$$h_e = \left[\frac{1 - C_e^2}{C_e^2}\right]VP_h$$

Where:

h_e = the **Hood Entry Losses**, and is as defined for Equation 4.9;

C_e = the **Coefficient of Entry** for the hood under consideration, and is also as defined for Equation 4.10; and

VP_h = the **Velocity Pressure** in the hood, which is measured in inches of water.

Equation 4.13

The following Equation 4.13 is known as the *Hood Entry Loss Factor Equation*, and it, too, is widely used.

$$F_h = \frac{h_e}{VP_h}$$

Where: F_h = the **Hood Entry Loss Factor**, which is a dimensionless parameter;

h_e = the **Hood Entry Losses**, and is as defined for Equation 4.9;

VP_h = is the **Velocity Pressure** in the hood, and is as defined for Equation 4.12, on the previous page, namely.

Equation 4.14

The following two variations of Equation 4.14, identified as Equations 4.14A and 4.14B, are known jointly as the *Compound Hood Equations*. They provide a vehicle for calculating the **Hood Static Pressure** for all conditions in which there are quantitative differences in the **Duct Velocity** and the **Hood Slot Velocity**. There are two specific circumstances that will be considered:

Condition 1: the **Duct Velocity** is *greater* than the **Hood Slot Velocity** (Equation 4.14A); and

Condition 2: the **Hood Slot Velocity** is *greater* than the **Duct Velocity** (Equation 4.14B).

In addition to these two *Compound Hood Equations*, there is an additional pair of relationships — actually approximations — that provide useful estimates as to the magnitudes of the **Entry Losses** in either the **Hood Slots** or the **Hood Ducts** of any compound hood. These two approximations have been designated as Equations 4.14C and 4.14D. These two Entry Losses are be included as factors in both Equations 4.14A and 4.14B.

Equation 4.14A — (Condition 1)

$$SP_h = h_{ES} + h_{ED} + VP_d \quad \text{whenever: } V_d > V_s$$

Where: SP_h = the **Hood Static Pressure**, and is as defined for Equation 4.10;

h_{ES} = the **Hood Slot Entry Loss**, measured in inches of water;

h_{ED} = the **Hood Duct Entry Loss**, measured in inches of water;

VP_d = the **Duct Velocity Pressure**, is as defined for Equation 4.9, and is measured in inches of water;

$$V_s \quad = \quad \text{the \textbf{Slot Velocity}, and is measured in velocity units, such as feet per minute (fpm); and}$$

$$V_d \quad = \quad \text{the \textbf{Duct Velocity}, which is measured in the same velocity units as } V_s.$$

Equation 4.14B — (Condition 2)

$$SP_h = h_{ES} + h_{ED} + VP_s \qquad \text{whenever: } V_s > V_d$$

Where: SP_h, h_{ES}, h_{ED}, V_s, and V_d are all exactly as defined for Equation 4.14A; and

$$VP_s \quad = \quad \text{the \textbf{Slot Velocity Pressure}, and is measured in inches of water.}$$

Note: For virtually any hood and/or duct situation, the following two approximations will always hold:

Equation 4.14C

$$h_{ES} = 1.78[VP_s]$$

Where: h_{ES} and VP_s are exactly as defined for Equations 4.14A and 4.14B.

Equation 4.14D

$$h_{ED} = 0.25[VP_d]$$

Where: h_{ED} and VP_d are exactly as defined for Equations 4.14A and 4.14B.

Calculations Involving the Rotational Speeds of Fans
Equation 4.15

Equation 4.15 relates the **Air Discharge Volume** being provided by a fan to its **Rotational Speed**. This is very important whenever one is trying to decide on the relative sizes of the pulleys that will be used (i.e., a pulley on the fan itself, and a second one on the motor that is to serve as the motive force for the fan). This relationship, as well as each of the five that will follow, assumes that the fan being evaluated may well be used under **different** operating circumstances, but only while handling fluids or gases of the **same** density.

$$\frac{CFM_1}{CFM_2} = \frac{RPM_1}{RPM_2}$$

Where: **CFM$_i$**   =   the **Air Discharge Volume** of the fan when it is operating at the ith set of conditions, in cubic feet per minute (cfm); and

        RPM$_i$ =   the ith operating **Rotational Speed** of the fan, in revolutions per minute (rpm).

Equation 4.16

Equation 4.16 relates the **Static Discharge Pressure** of a fan to its **Rotational Speed**. As with Equation 4.15, this relationship assumes that the fan will be evaluated *only* for gases or fluids of the *same density*.

$$\frac{SP_{Fan_1}}{SP_{Fan_2}} = \left[\frac{RPM_1}{RPM_2}\right]^2$$

Where: SP_{Fan_i} =   the **Static Discharge Pressure** of a fan when it is operating at the ith set of conditions; and

        RPM$_i$ =   is the ith operating **Rotational Speed** of the fan, as defined for Equation 4.15.

Equation 4.17

Equation 4.17 relates the motive **Brake Horsepower** required to operate a fan at any of its possible **Rotational Speeds**. As with the two previous equations, this relationship assumes that the fan will be evaluated only for different **Rotational Speed** applications that involve gases and fluids having the *same density*.

$$\frac{BHP_1}{BHP_2} = \left[\frac{RPM_1}{RPM_2}\right]^3$$

Where: **BHP$_i$**   =   the **Brake Horsepower** required for a fan to be operated at the ith set of conditions, measured in brake horsepower units (bhp); and

        RPM$_i$ =   is the ith operating **Rotational Speed** of the fan, as defined in Equation 4.15.

Calculations Involving the Diameters of Fans
Equation 4.18

Equation 4.18 relates the **Air Discharge Volume** being produced by a fan to its **Diameter**. This relationship, like the previous three, as well as each of the two that will follow, assumes that the fan being examined will only be evaluated for *different diameters*, while handling gases or fluids of the *same density*.

$$\frac{CFM_1}{CFM_2} = \left[\frac{d_1}{d_2}\right]^3$$

Where: CFM_i =   the **Air Discharge Volume** of the fan at its ith **Diameter**, this **Air Discharge Volume** is measured in cubic feet per minute (cfm); and

d_i =   the **Diameter** of the ith Fan, measured in inches.

Equation 4.19

Equation 4.19 relates the **Static Discharge Pressure** being developed by a fan to its **Diameter**. This relationship, like the previous four, assumes that the fan being evaluated will only be operated at *different diameters* while handling gases and fluids of the *same density*.

$$\frac{SP_{Fan_1}}{SP_{Fan_2}} = \left[\frac{d_1}{d_2}\right]^2$$

Where: SP_{Fan_i} =   the **Static Discharge Pressure** of a fan of the ith **Diameter**, this **Static Discharge Pressure** is measured in inches of water; and

d_i =   the **Diameter** of the ith Fan, as defined for Equation 4.18.

Equation 4.20

Equation 4.20 relates the **Brake Horsepower** required to operate Fans of various **Diameters**. This relationship — as was true with the previous five — assumes that the fan under evaluation will only be operated at *different diameters* while handling gases or fluids of the *same density*, and at the *same* **Fan Discharge Volume**.

$$\frac{BHP_1}{BHP_2} = \left[\frac{d_1}{d_2}\right]^5$$

Where: BHP_i =   the **Brake Horsepower** required for a fan of the ith **Diameter** to deliver some   specified,   and   for   this application, identical **Air Discharge Volume**. The **Brake Horsepower** term is always measured in brake horsepower units (bhp); and

d_i = the **Diameter** of the ith fan, as defined for Equation 4.18.

Calculations Involving Various Other Fan-Related Factors

Equation 4.21

Equation 4.21 is commonly known as the *Fan Brake Horsepower Equation*. It combines most of the factors that must be considered in the choice of a fan for any ventilation application.

$$BHP = \frac{1}{63.56}\left[\frac{(CFM)(TP)}{FME}\right] = 0.0157\left[\frac{(CFM)(TP)}{FME}\right]$$

Where: **BHP** = the **Brake Horsepower** that will be required for the fan to be able to provide the required performance level, measured in brake horsepower units (bhp);

CFM = the required **Fan Discharge Volume**, measured in cubic feet per minute (cfm);

TP = the **Fan Total Pressure**, measured in inches of water; and

FME = the **Fan Mechanical Efficiency**, which is a dimensionless percentage ≤ 100%.

Equation 4.22

Equation 4.22 is known as the *Fan Total Pressure Equation*. It is a very simple, straightforward, and widely used relationship.

$$TP_{Fan} = TP_{out} - TP_{in}$$

Where: TP_{Fan} = the **Fan Total Pressure**, measured in inches of water;

TP_{out} = the **Fan Total Output Pressure**, measured in inches of water; and

$$TP_{in} = \text{the \textbf{Fan Total Input Pressure},}$$
measured in inches of water.

Equation 4.23

Equation 4.23 is known as the *Fan Static Pressure Equation*, simply because it defines the overall **Fan Static Pressure**.

$$SP_{Fan} = SP_{out} - SP_{in} - VP_{in}$$

Where: SP_{Fan} =   the **Fan Static Discharge Pressure**, measured in inches of water;

SP_{out} =   the **Fan Outlet Static Pressure**, measured in inches of water;

SP_{in} =   the **Fan Inlet Static Pressure**, measured in inches of water; and

VP_{in} =   the **Fan Inlet Velocity Pressure**, also measured in inches of water.

Calculations Involving Air Flow Balancing at a Duct Junction

Equation 4.24

Equation 4.24A, and the family of relationships that follow it, define completely the process of balancing a duct junction. This set makes up the *Duct Junction Balancing System*. This system is logical in its development, and in the solution path it implies. Basically, this procedure considers the volumetric flow rate in each of two ducts that join to form a single larger duct.

Equation 4.24A

$$R = \frac{SP_{greater}}{SP_{lesser}}$$

Where: **R**   =   the **Junction Balance Ratio**, which is a dimensionless number;

$SP_{greater}$ =   the **Duct Static Pressure** in the duct that carries the greater flow volume into the duct junction, measured in inches of water; and

SP_{lesser} =   the **Duct Static Pressure** in the duct carrying the lesser flow volume into the duct junction, also measured in inches of water.

We must now consider *Three* Different Scenarios:

1. **R < 1.05** Consider the duct junction to be balanced.
2. **R > 1.20** The duct junction is so very badly *imbalanced* that the entire ventilation system must be *completely redesigned* from the ground up.
3. **1.05 ≤ R ≤ 1.20** Increase the volumetric flow rate in the lesser volume branch from Q_{former} to Q_{new}, according to the following relationship, identified as Equation 4.24B.

Equation 4.24B

$$Q_{new} = Q_{former} \sqrt{\frac{SP_{greater}}{SP_{lesser}}} = Q_{former} \sqrt{R}$$

Where: Q_{new} = the *increased* flow rate in the branch in which the lesser volume has been flowing, measured in cubic feet/minute (cfm);

Q_{former} = the *starting* flow rate in the branch in which the lesser volume has been flowing, measured in the same units as Q_{new}.

$SP_{greater}$ = the **Duct Static Pressure** in the duct that carries the greater flow volume into the duct junction, as defined for Equation 4.24A, measured in inches of water;

SP_{lesser} = the **Duct Static Pressure** in the duct carrying the lesser flow volume into the duct junction, also as defined for Equation 4.24A, and also measured in inches of water; and

R = the **Junction Balance Ratio**, which is a dimensionless number, as defined for Equation 4.24A, and also measured in inches of water.

Calculations Involving Dilution Ventilation
Equation 4.25

Equation 4.25 provides a single relationship that makes it possible to calculate the steady state, **Equilibrium Concentration** that would be produced in a room — or any enclosed space for which the overall volume can be determined — by the complete evaporation of some specific volume of any identifiable volatile solvent. This is probably the single most complicated equation with which any professional will ever have to work — its principal saving grace is that it works. This relationship assumes Normal Temperature and Pressure (NTP).

$$C = 6.24 \times 10^7 \left[\frac{V_s \rho T}{P_{atm} V_{room} (MW_s)} \right]$$

Where:

C = the **Equilibrium Concentration** of the volatile solvent that would be produced in the room by the evaporation of the known volume of solvent, measured in ppm(vol);

V_s = the liquid **Volume** of the solvent that has evaporated, measured in milliliters (ml);

ρ = the **Density** of the solvent, measured in grams per cubic centimeter (gms/cm^3);

T = the absolute **Temperature** in the room, measured in degrees Kelvin (K);

MW_s = the **Molecular Weight** of the solvent;

P_{atm} = the **Ambient Barometric Pressure** that is prevailing in the room, measured in millimeters of mercury (mm Hg);

V_{room} = the **Volume** of the Room, measured in liters; and

6.24×10^7 = the **Proportionality Constant** that makes this equation valid under NTP conditions.

Equation 4.26

Equation 4.26 is known as the *Basic Room Purge Equation*. It provides the necessary relationship for determining the time required to reduce a known initial high level concentration of *any* vapor — existing in a defined closed space or room — to a more acceptable ending lower level concentration. The relationship has been provided in two formats: one employing natural logarithms and the other, common logarithms.

$$D_t = \left[\frac{V}{Q}\right] \ln\left[\frac{C_{initial}}{C_{ending}}\right] \quad — \text{Natural Log Format}$$

$$D_t = 2.3\left[\frac{V}{Q}\right] \log\left[\frac{C_{initial}}{C_{ending}}\right] \quad — \text{Common Log Format}$$

Where:

D_t = the **Time Required** to reduce the vapor concentration in the closed space or room, as required, measured in minutes;

V = the **Volume** of the closed space or room, measured in cubic feet (ft^3);

Q = the **Ventilation Rate** at which the closed space or room will be purged by whatever air handling system is available for that purpose, measured in cubic feet per minute (cfm);

$C_{initial}$ = the **Initial High Level Concentration** of the vapor in the ambient air of the closed space or room, which concentration is to be reduced — by purging at Q cfm — to a more acceptable ending lower level concentration, measured in ppm(vol); and

C_{ending} = the desired **Ending Lower Level Concentration** of the vapor that is to result from the purging effort in the closed space or room, also measured in ppm(vol).

Equation 4.27

Equation 4.27 is known as the *Purge-Dilution Equation*. It is the most basic and fundamental relationship available for determining the various parameters associated with reducing the concentration of the vapor of any volatile material in a closed space or room.

$$C = C_0 e^{-\left[V_{removed}/V_{room}\right]}$$

Where:

C = the **Ending Concentration** of the vapor in the closed space or room,

$$\text{which concentration, measured in ppm (vol), resulted from the purging activities;}$$

C_o = the **Initial Concentration** of the vapor in the closed space or room that is to be reduced by purging, also measured in ppm(vol);

$V_{removed}$ = the **Air Volume** that has been withdrawn from the closed space or room, measured in *any* suitable volumetric units, usually in cubic feet (ft^3); and

V_{room} = the **Volume** of the room, measured in the *same* volumetric units as $V_{removed}$, which is usually in cubic feet (ft^3).

Chapter 5
Thermal Stress

Thermal stress can arise from exposures to either hot or cold environments. For a wide variety of work environments it is common for there to be extremes as well as wide fluctuations in the temperature; because of this, it is important to understand thermal stress. This chapter will focus on the factors and parameters that are used to characterize thermal stress, and, in addition, will discuss the effects that these conditions can produce on individuals who, in the normal course of their work, may be exposed to them.

Relevant Definitions
Thermal Stress
Heat Stress

Heat Stress is a condition that arises from a variety of factors among the most important of which are:

1. the ambient temperature,
2. the relative humidity,
3. the level of effort required by the job, and
4. the clothing being worn by an exposed individual.

An individual who is experiencing **Heat Stress** will tend to exhibit an array of measurable symptoms which can include some or all of the following:

1. an increased pulse rate,
2. a greater rate of perspiration, and
3. an increase in the individual's body temperature.

Heat Stress Disorders

The five physical disorders that can arise from heat stress, listed in increasing order of severity, are as follows:

1. *Heat Rash*	A heat rash — also often referred to as "prickly heat" — tends to arise in an individual after a period of prolonged sweating. It is characterized by an itchy reddening of the skin and a sudden decrease in the rate of perspiration.
2. *Cramps*	Heat cramps arise as a result of prolonged periods of a combination of sweating and a lack of fluid and salt intake. Such a situation causes an overall body electrolyte imbalance, and the primary symptomatic manifestation is severe muscle cramps, most frequently in the abdomen.

77

3. *Dehydration*	Dehydration is the result of excessive fluid loss. Among its most common causes are: excessive sweating, vomiting, diarrhea, and/or alcohol consumption. Symptoms of dehydration are often subtle but include exhaustion, overall weakness, dry mouth, decreased work output, etc.
4. *Heat Exhaustion*	Heat exhaustion arises from extreme cases of dehydration. It is characterized by some or all of the following conditions or symptoms: increasing pulse rate, decreasing blood pressure, slight to moderate increases in body temperature, fatigue, increasing levels of sweating, lack of skin color, dizziness, blurred vision, headache, decreased work output, and collapse.
5. *Heat Stroke*	Heat stroke is usually the result of very significant overexposure to the factors of heat stress. It can also arise from drug or alcohol abuse and on occasion from genetic factors. Heat stroke is almost always accompanied by an increase in body temperature to levels greater than 104°F (40°C). Symptomatic indications include: chills, irritability, hot and dry skin, convulsions, and unconsciousness.

Cold Stress

Cold Stress differs dramatically from Heat Stress. Typically a body will adapt to conditions of Heat Stress by increasing its level of perspiration in an effort to provide increased cooling. **Cold Stress** adaptations usually involve a decrease in the blood flow to the skin and the extremities. The principal causes of **Cold Stress** are exposure to cold temperatures and vibrations, either singly or in combination.

Cold Stress Disorders

The four physical disorders that can arise from cold stress, listed in increasing order of severity, are as follows:

1. *Chilblains*	Chilblains usually arise as a result of inadequate clothing during periods of exposure to cold temperatures and high relative humidities. Reddening of the skin accompanied by localized itching and swelling are the principal indications of chilblains.
2. *Frostnip*	Frostnip, which is similar to frostbite, results from prolonged, unprotected exposures to cold temperatures above 32°F (0°C). Symptoms of frostnip are areas of pain and/or itching, and a distinct whitening of the skin.

3. *Frostbite*   Frostbite is produced from unprotected exposures to cold temperatures at or below freezing — i.e., ≤ 32°F or 0°C. Frostbite is characterized by the sequential change in skin color from white to gray to black (depending upon the temperature and the length of exposure), a reduction in the sensations of touch ranging from slight to total (again depending upon the temperature and the length of exposure), and numbness.

4. *Hypothermia*   Hypothermia results from extreme exposures to the factors of cold stress, coupled possibly with dehydration and/or exhaustion. Alcohol and/or drug abuse also can contribute to hypothermia. A person who is experiencing hypothermia will usually show some or all of the following symptoms: chills, euphoria, pain in the extremities, slow and weak pulse, body temperature of less than 95°F (35°C), fatigue, drowsiness, and unconsciousness.

Ambient Conditions Related to Thermal Stress
Dry Bulb Temperature

The **Dry Bulb Temperature** is the most direct measurement of air temperature. By definition, it is to be accomplished by the use of a capillary thermometer that is completely exposed to and/or immersed in the air mass whose temperature is to be measured. This thermometer should be shielded from sources of radiant heat.

Wet Bulb Temperature

The **Wet Bulb Temperature** of an air mass differs from the Dry Bulb Temperature measurement by the fact that the fluid reservoir bulb of the capillary thermometer that is used to make this measurement is encased in a sheath of water moistened cloth. This wet sheath provides cooling to the thermometer bulb by the evaporation of water, in most cases causing the **Wet Bulb Temperature** to be less than its Dry Bulb Temperature counterpart — the obvious exception to this is the case where the ambient relative humidity is 100%, a condition wherein evaporation, and the resulting evaporative cooling, are effectively eliminated.

There are actually two categories of **Wet Bulb Temperature**. The first is the **Natural Wet Bulb Temperature** which is obtained simply by encasing a capillary thermometer bulb in a wet cloth and then using this combination to make an air temperature measurement. The other category is described as a **Psychrometric Wet Bulb Temperature. Psychrometric Wet Bulb Temperatures** are obtained by the use of a sling psychrometer, a tool that is made up of a pair of identical capillary thermometers, one of which is bare while the other is sheathed in a wet cloth. To obtain a reading from a sling psychrometer, this mechanism is whirled through the air, a process that produces a maximized rate of evaporative cooling for the wet bulb. The difference in the temperatures indicated by the two thermometers of a sling psychrometer can then be used to determine the relative humidity of the air mass being measured.

Air or Wind Speed

The **Air** or **Wind Speed** is simply the rate at which a mass of air is passing an arbitrary stationary point. The direction of movement is not important since this measure is used principally in a determination of the convective heat transfer to and from the air. It is typically measured by an anemometer.

Globe Temperature

The **Globe Temperature** of an air mass arises from the combination of heat input by radiation from the surroundings coupled with the simultaneous heat loss by the convective movement of air around the **Globe Temperature** measurement device, which is a six-inch diameter, thin-walled copper, spherical globe, painted matte black with an appropriate temperature sensor at its center.

Effective Temperature

The **Effective Temperature** is an index that is used to relate the subjective effect that the thermal environment might be expected to have on the comfort of an individual who is exposed to that environment. It is a combination of the Dry Bulb, the Wet Bulb, and the Globe Temperatures.

Wet Bulb Globe Temperature Index

The **Wet Bulb Globe Temperature Index** (abbreviated, **WBGT**) is the most widely used algebraic approximation of an "Effective Temperature" currently in use today. It is an Index that can be determined quickly, requiring a minimum of effort and operator skill. As an approximation to an "effective temperature", the **WBGT** takes into account virtually all the commonly accepted mechanisms of heat transfer (i.e., radiant, evaporative, etc.). It does not account for the cooling effect of wind speed. Because of its simplicity, **WBGT** has been adopted by the American Conference of Government Hygienists (ACGIH) as its principal index for use in specifying a heat stress related Threshold Limit Value (**TLV**). For outdoor use (i.e., in sunshine), the **WBGT** is computed according to the algebraic sum:

$$\mathbf{WBGT} = 0.7 \ (NWB) + 0.2 \ (GT) + 0.1 \ (DB).$$

For indoor use, the **WBGT** is computed according to the following slightly modified algebraic sum:

$$\mathbf{WBGT} = 0.7 \ (NWB) + 0.3 \ (GT).$$

where: (NWB) =   Natural Wet Bulb Temperature,

(GT)   =   Globe Temperature, and

(DB)   =   Dry Bulb Temperature.

Relevant Formulae and Relationships
Thermal Stress (Indoor/Outdoor),
With and Without Solar Load

Equation 5.1

Equation 5.1 is the relationship that provides the **Wet Bulb Globe Temperature Index (WBGT)** that is applicable only to situations for which there is *no solar load* (i.e., no direct solar input to the condition or circumstance of the area or space being evaluated). Obviously, most indoor situations fulfill this requirement; in addition, any outdoor circumstance wherein the sun has been shaded — i.e., where it makes no radiant contribution to the thermal or temperature environment — also fulfills this condition. This category of the **Wet Bulb Globe Temperature Index** is usually identified with an "$_{Inside}$" subscript.

$$WBGT_{Inside} = 0.7[NWB] + 0.3[GT]$$

Where: $WBGT_{Inside}$ = the **Wet Bulb Globe Temperature Index**, applicable to any situation in which there is no Solar Load, usually measured in °C;

NWB = the **Natural Wet-Bulb Temperature**, usually also measured in °C; however, it should be noted that any temperature scale may be used for these parameters, so long as the units of every temperature parameter in the formula are consistent with the units of every other temperature parameter; and

GT = the **Globe Temperature**, also in units consistent with every other parameter in this formula.

Equation 5.2

Equation 5.2 is the alternative relationship that provides the **Wet Bulb Globe Temperature Index (WBGT)** that is applicable to situations for which there is a measurable solar load. Outdoor conditions usually require this approach; and, correspondingly, this category of the **Wet Bulb Globe Temperature Index** is usually identified with an "$_{Outside}$" subscript.

$$WBGT_{Outside} = 0.7[NWB] + 0.2[GT] + 0.1[DB]$$

Where: $WBGT_{Outside}$ is precisely as defined, under the designation of $WBGT_{Inside}$;

NWB is as defined previously;

GT   is as defined previously; and

DB = the **Dry-Bulb Temperature**, usually measured in °C; however, it should be noted for this relationship, too, that any temperature scale may be used for these parameters, so long as the units of every temperature parameter in the equation are consistent with the units of every other temperature parameter.

Temperature Related Time Weighted Averages

Equation 5.3

Equation 5.3 provides the relationship necessary for the determination of the **Effective Time Weighted Average WBGT Index**. This Index represents an average exposure over time, and at various different **WBGT Indices**. This formula is directly analogous to every other formula that is used to determine a Time Weighted Average.

$$
\text{WBG } T_{TWA} = \frac{\displaystyle\sum_{i=1}^{n} [WBGT_i][t_i]}{\displaystyle\sum_{i=1}^{n} t_i} =
$$

$$
\frac{[WBGT_1][t_1] + [WBGT_2][t_2] + \ldots + [WBGT_n][t_n]}{t_1 + t_2 + \ldots + t_n}
$$

Where: **WBGT$_{TWA}$** = the **Effective Time Weighted Average WBGT Index** that corresponds to a varied thermal exposure over time, usually measured in °C or °F;

WBGT$_i$ = the ith **Wet Bulb Globe Temperature Index** that was measured over the Time Interval, t_i, usually measured in °C or °F (either "WBGT$_{Indoor}$" or "WBGT$_{Outdoor}$", but *never* a mixture of "WBGT$_{Indoor}$" and "WBGT$_{Outdoor}$"); and

t_i = the ith **Time Interval**, usually measured in hours, can be measured in any useful and consistent units.

Chapter 6
Sound and Noise

The general interest in this topic stems from the steadily growing incidence in today's modern industrial workplace, of sound and/or noise-induced hearing impairments (mostly partial, but occasionally, extending even to total hearing loss). The differences between sounds and noise are subjective. Sounds are considered to be something pleasant or useful — such as music (pleasant) or speech (useful). Noise, in contrast, is thought of as being unpleasant, consisting of such things as the sounds of a table saw cutting wood or a fingernail scratching a chalkboard. This chapter will focus on the factors, parameters, and relationships that permit an accurate assessment of the potential for physiological damage caused by the ambient noise levels that exist in the workplace.

Relevant Definitions
Categories of Noise

Continuous Noise

An unbroken sound, made up of one or more different frequencies of either constant or varying sound intensity/sound pressure level, is referred to as **Continuous Noise**. If such a sound is constant and unvarying in its amplitude, it would be referred to as "steady" **Continuous Noise**. The alternative to this would be "varying" **Continuous Noise**. **Continuous Noise** is a fairly common occurring phenomenon — both in the industrial and the natural environment.

In the natural environment, one might regard the sound of a waterfall as "steady" **Continuous Noise**, while the sounds of wind blowing through a forest would be in the "varying" category.

In the industrial environment, the sound of a rotating electric motor (i.e., a fan, a pump, etc.) would be "steady" **Continuous Noise**, while the operation of a floor waxer, relative to a fixed observer, would be "varying" **Continuous Noise**.

Continuous Noise is an extremely useful concept. In assessing the potential hazards of any noise filled environment, one attempts to quantify the existing noise pattern in terms of the "steady" **Continuous Noise** that could, in theory, replace it without altering any of the adverse effects that might be being experienced by a human observer. For any environment, an $L_{equivalent}$, or the "steady" equivalent **Continuous Noise** level, as described above, can usually be determined; the actual sound intensity level of this "steady" **Continuous Noise** — which equals $L_{equivalent}$ —then can be used to evaluate the overall sound hazard that is posed to individuals who must occupy that environment.

Intermittent Noise

Intermittent Noise is a broken or noncontinuous sound (i.e., sound bursts or periods of time during which there are intervals of quiet (nonsound) and subsequent intervals during which there is measurable sound). **Intermittent Noise** can also be made up of one or more different frequencies of sound, of either constant or varying intensity or sound level. The sound of an operating typewriter would be considered an intermittent noise.

Although such a category of noise is more difficult to relate to in the context of an $L_{equivalent}$, such a determination can be, and frequently is, made simply by integrating, over time, the entirety of the noise regime in any setting. Most commercially available sound level meters have the capability to provide an $L_{equivalent}$ for any situation in which sound is to be measured, whether this ambient sound or noise is intermittent, continuous, or a mixture of these different categories.

Characteristics of Sounds and/or Noise

Frequency

The **Frequency** of any sound or noise is the time rate at which complete cycles of high and low pressure regions (compressions and rerefactions) are produced by the noise or sound source. The most common unit of sound **Frequency** is the Hertz (abbreviated, Hz), which is the number of complete cycles that occur in a period of one second. The frequency range over which the human ear can hear varies with age and circumstance; however, a normal hearing "young" ear will usually be able to distinguish sounds, at moderate levels, in the range 20 to 20,000 Hz = 20 to 20,000 cycles/second.

Frequency Band

A sound can be made up of a single frequency — i.e., a tuning fork set at middle "C"; however, it is far more common for a sound or a noise, coming from any source, to be made up of a combination of different frequencies. Whenever a sound or a noise consists of a set of closely related frequencies, this set can be described as a **Frequency Band**. To identify any specific **Frequency Band**, one need only identify the range of frequencies that make it up, namely, by the lowest and the highest frequency in its "inventory." These two frequencies are known as the "Upper and Lower Band-Edge Frequencies" of the particular **Frequency Band** being described.

Octave Bands and Bandwidths

Probably the most commonly used Frequency Band would be the **Octave Band**. A typical **Octave Band** is always characterized by the single frequency numerically located at its geometric center. The "Center

Frequency" for any **Octave Band** is the *geometric mean* of its "Upper and Lower Band-Edge Frequencies."

A second important characteristic of an **Octave Band** is the range of frequencies in it, or its **Bandwidth**. The **Bandwidth** of any Frequency Band is the range between its "Upper and Lower Band-Edge Frequencies".

Typically, for a full **Octave Band**, this range will be up to one full octave in total **Bandwidth** — the principal characteristic of one full octave is that its highest frequency (the Upper Band-Edge Frequency) is always exactly twice its lowest frequency (the Lower Band-Edge Frequency). A tabulation of the standard single, or full, **Octave Bands** is shown below.

Full Octave Bands

Center Frequency	Lower Band-Edge Frequency	Upper Band-Edge Frequency
31.3 Hz	22.1 Hz	44.2 Hz
62.5 Hz	44.2 Hz	88.4 Hz
125.0 Hz	88.4 Hz	176.8 Hz
250.0 Hz	176.8 Hz	353.6 Hz
500.0 Hz	353.6 Hz	707.1 Hz
1,000.0 Hz	707.1 Hz	1,414.2 Hz
2,000.0 Hz	1,414.2 Hz	2,828.4 Hz
4,000.0 Hz	2,828.4 Hz	5,656.9 Hz
8,000.0 Hz	5,656.9 Hz	11,313.7 Hz
16,000.0 Hz	11,313.7 Hz	22,627.4 Hz

Sound Wavelength

The **Wavelength** of a **Sound** is the precise distance required for one complete pressure cycle (i.e., one cycle of high (compressed) and low (rarefied) pressure regions) for that frequency of sound. Since sound is a periodic wave phenomenon — even though a markedly different one than the more classic example, light — it can be characterized in terms of its wavelength. This is most easily recognized by considering such things as organ pipes, the lengths of which will always relate to the wavelength of the sound that the particular pipe produces.

Pitch

The **Pitch** of a sound is the subjective auditory perception of the frequency of that sound. It, of course, depends upon the sound frequency, but also on its waveform, on the number of harmonics or overtones present, and on the overall sound pressure level.

Velocity of Sound

The **Velocity of Sound** is the speed at which the sequential regions of high and low pressure propagate away from the source of the sound. For all practical purposes this velocity can be considered to be a constant through whatever medium the sound is transiting. It varies directly as the square root of the density of the medium involved, and inversely as the compressibility of that medium. For example:

Medium	Velocity of Sound
air	~ 1,130 ft/sec
sea water	~ 4,680 ft/sec
hard wood	~ 13,040 ft/sec
steel	~ 16,550 ft/sec

Loudness

The **Loudness** of a sound is an observer's impression of its amplitude. This subjective judgment is influenced strongly by the characteristics of the ear that is doing the hearing.

Characteristic Parameters of Sound and Noise
Sound Intensity and Sound Intensity Level

The **Sound Intensity** of any sound source at any particular location is the average rate at which sound energy from that source is being transmitted through a unit area that is normal to the direction in which the sound is propagating. The most common units of measure for Sound Intensity are joules per square meter (m^2) per second, which are also equal to units of watts per square meter (m^2).

Sound Intensity is usually expressed in terms of an appropriate **Sound Intensity Level**. This parameter is determined by ratioing the Sound Intensity of some noise/sound against the accepted reference base Sound Intensity, which is 10^{-12} watts/m^2. When determined in this manner (see Equation 6.3), the units of the Sound Intensity Level will be in decibels (dBs).

Sound Power and Sound Power Level

The **Sound Power** of any sound source is the total sound energy radiated by that source per unit time. The most common units of measure for Sound Power are watts.

Sound Power is usually expressed in terms of an appropriate **Sound Power Level**. This parameter can be calculated by ratioing the Sound Power of some noise/sound source against the accepted reference base Sound Power, which is 10^{-12} watts. When determined in this manner (see Equation 6.4), the units of the Sound Power Level will also be in decibels (dBs).

Sound Pressure

Sound Pressure normally refers to the RMS values of the pressure changes, above and below atmospheric pressure, which are used to measure steady state or continuous noise. The three most common units of measure for **Sound Pressure** are:

$$\text{newtons per square meter} = n/m^2 = \text{pascals}$$
$$(1 \ n/m^2 = 1 \ Pa)$$

$$\text{dynes per square centimeter} = d/cm^2$$
$$\text{microbars}$$

Root-Mean-Square (RMS) Sound Pressure

The **Root-Mean-Square (RMS)** value of any changing quantity, such as sound pressure, is equal to the square root of the mean of the squares of all the measured instantaneous values of that quantity.

Common Measurements of Sound Levels
Sound Pressure Level

Evaluation and/or measurement of any sound or noise, from the perspective of the characteristics of the healthy human ear, poses a difficult problem. This problem arises because of the very wide range of Sound Pressures that the human ear can hear without incurring damage. The healthy human ear can detect sounds at extremely low Sound Pressures (i.e., $L_p = 20 \ \mu Pa$), and can survive without damage sounds having very high Sound Pressures (i.e., $L_p = 2 \times 10^8 \ \mu Pa = 200 \ Pa$).

When evaluated at a reference frequency of 1,000 Hz, the "effective operating range" of a healthy human ear involves 7+ orders of magnitude of actual Sound Pressures. Because these "human hearing related" significant Sound Pressures vary over such an extremely wide range, the parameter that is most commonly used to describe Sound Pressures is the **Sound Pressure Level** (the **SPL**). This parameter can be determined by ratioing the measured Sound Pressure of some noise or sound source against the reference base Sound Pressure of $2 \times 10^{-5} \ n/m^2 = 2 \times 10^{-5} \ Pa = 20 \ \mu Pa$. When determined in this manner (see Equation 6.2), the units of the Sound Pressure Level will also be expressed in decibels (dBs). The decibel was chosen in this situation simply because logarithmic units of measure are virtually always judged to be more useful when dealing with parameters whose values may vary over more than 4 or 5 orders of magnitude.

Threshold of Hearing

The **Threshold of Hearing** for a healthy human ear, expressed in decibels and determined at a frequency of 1,000 Hz, is **0 dB**. This is the approximate sound of a feather falling in an otherwise completely quiet room — it is doubtful that the frequency of sound produced by a falling feather would be 1,000 Hz; however, the author can find no other reference to any physical event that would produce this low level of an SPL.

Threshold of Pain

The **Threshold of Pain** for a healthy human ear, also expressed in decibels and also determined at a frequency of 1,000 Hz, is approximately **140 to 145 dB**. At this SPL, an exposed individual would likely experience both permanent damage to his or her hearing and in addition experience actual pain, thus the name. The sound of a commercial jet plane taking off, 25 feet from the unprotected observer, would produce this approximate SPL.

Sound and/or Noise Measurement Time Weightings

In the quantification of various Sound Pressure Levels (in decibels), there are four different commonly used averaging periods, or **Time Weightings**, that are part of the standard RMS detection method. These four are: Peak, Impulse, Fast, and Slow Noise Weightings.

Peak Noise

A burst of sound having a duration of less than 100 milliseconds is considered to be in the **Peak Noise** category. Such sounds will also fall under the category of Impulsive or Impact Noise.

Impulsive or Impact Noise

The types of noise produced by such things as a gun being fired, the operation of an industrial punch press, or the use of a hammer to drive a nail are all highly transient sound phenomena, and are usually treated as **Impulsive or Impact Noises**. This type of noise is defined to be any sound having an amplitude rise time of 35 milliseconds or less, and a fall time of 1,500 milliseconds or less.

Fast Time Weighted Noise

Sound pressure level measurements using a 125-millisecond moving average time weighting period are said to have been determined using **Fast Time Weighting**.

Slow Time Weighted Noise

Sound pressure level measurements using a 1.0-second moving average time weighting period are said to have been determined using **Slow Time Weighting**.

Sound and/or Noise Measurement Frequency Weightings
Linear Frequency Weighting

Any measurement of a sound pressure level can be thought of as the unique "sum" of the 10 discrete sound pressure levels of the standard Octave Bands that have made up the sound being monitored. If these measurements are developed without the application of any internal "adjustments" by the sound level meter that is being used — i.e., the meter neither increases nor decreases its measured decibel level of any of the Octave Bands before developing its overall "sum" measurement — then the result is said to have been produced using **Linear Frequency Weighting**. Whenever one attempts to characterize a noise, for the purpose of designing or implementing some sort of sound mitigation, the measurements will probably employ a Linear Weighting approach. If it is desired to measure any one single Octave Band, **Linear Frequency Weighting** will always be employed.

A-Frequency Weighting Scale

The **A-Frequency Weighting Scale** (covered in complete quantitative detail in the next subsection of this Chapter, namely, **Relevant Formulae and Relationships**, as the first part of Equation 6.10) is a set of measurement weightings that must be applied to the decibel reading for each of the standard Octave Bands that make up the sound being measured. The application of these weighted adjustments — by the internal A-Weighting network in the sound level meter that is making the measurement — ensures that the resultant indicated overall Sound Pressure Level measurement will be of a magnitude that constitutes the very best approximation to what a normal human ear would have perceived. The A-Weighting Scale is usually thought to apply to noises having relatively low level intensities. Sound Pressure Level measurements made using this weighting are always identified by the inclusion of the letter "A" after the "dB" unit; thus, **dBA**. The **A-Frequency Weighting Scale** is the most commonly used and widely accepted frequency weighting scale employed in sound pressure level measurements today.

B-Frequency Weighting Scale

The **B-Frequency Weighting Scale** (also covered in complete quantitative detail in the next sub-section of this chapter, namely, **Relevant Formulae and Relationships**, as the second part of Equation 6.10) is one of the two alternative sets of measurement weightings that must also be applied to each of the standard Octave Bands that make up any sound. The application of these particular weightings — again by a sound level meter's internal B-Weighting network — is designed to produce a result that approximates what a normal human ear would have perceived to noises having relatively moderate, in contrast to low, intensities. Sound Pressure Level measurements made using this category of weighting are always identified by the inclusion of the letter "B" after the dB unit, thus, **dBB**. The **B-Frequency Weighting Scale** is not in very wide use today.

C-Frequency Weighting Scale

The **C-Frequency Weighting Scale** (also covered in complete quantitative detail in the next sub-section of this Chapter, namely, **Relevant Formulae and Relationships**, as the third and final part of Equation 6.12) is the second of the two alternative sets of measurement weightings that must also be applied to each of the standard Octave Bands that make up any sound. The application of these particular weightings — again by a sound level meter's internal C-Weighting network — is designed to produce a measurement that approximates what a normal human ear would have perceived as high intensity noises. Sound Pressure Level measurements made using this category of weighting are always identified by the inclusion of the letter "C" after the dB unit, thus, **dBC**. The **C-Frequency Weighting Scale** is only rarely used today.

Relevant Formulae and Relationships
Approximate Velocity of Sound in Air

The velocity of sound in the earth's atmosphere varies directly as the square root of the density of the air. The most easily measured parameter that affects the velocity of sound in the air is its prevailing ambient temperature.

Equation 6.1

The following relationship, Equation 6.1, was empirically derived; however, it has proven to be very accurate for calculating the **Velocity of Sound in Air** over a very wide range of ambient temperatures.

$$V = 49\sqrt{t + 459}$$

Where: V =   the Velocity of Sound in Air, measured in feet per second; and

t =   the Ambient Air Temperature, measured in relative English Units, namely, °F.

Basic Sound Measurements — Definitions
Equation 6.2

The following relationship, Equation 6.2, constitutes the definition of a **Sound Pressure Level**. The expression relates the measured Analog Sound Pressure Level to a "Base Reference Analog Sound Pressure Level", defining the common logarithm of this ratio to be the **Sound Pressure Level**. Because of the potential for extremely wide variations in the measurable analog Sound Pressures, the unit of measure for the **Sound Pressure Level** is the decibel, which, as stated above, is logarithmic and, as such, is better suited as a measure of numeric values, the magnitude of which can vary over several orders of magnitude.

$$L_P = 20\log\left[\frac{P}{P_0}\right] = 20\log\left[\frac{P}{2 \times 10^{-5}}\right] = 20\log P + 93.98$$

Where: L_P =   the **Sound Pressure Level**, measured in decibels (dBs);

                       P =   the measured Analog Sound Pressure Level, in units of newtons/square meter (nt/m^2); and

                       P_0 =   the "Base Reference Analog Sound Pressure Level", which has been defined to have a value of 2×10^{-5} nt/m^2.

Equation 6.3

This relationship, Equation 6.3, is the definition of a **Sound Intensity Level**. In this case, the equation relates the measured Analog Sound Intensity Level to a "Base Reference Analog Sound Intensity Level". As was true in the preceding case, this parameter also is measured in units of decibels, and again for the very same reasons.

$$L_I = 10\log\left[\frac{I}{I_0}\right] = 10\log\left[\frac{I}{10^{-12}}\right] = 10\log I + 120$$

Where: L_I =   the **Sound Intensity Level**, measured in decibels (dBs);

                       I =   the Analog Sound Intensity Level, measured in watts/square meter (wts/m^2); and

                       I_0 =   the "Base Reference Analog Sound Intensity Level", which has been defined to have a value of 10^{-12} wts/m^2.

Equation 6.4

This final analogous relationship, Equation 6.4, constitutes the definition of a **Sound Power Level**. As is the case for its two previous close relatives, this expression relates the measured Analog Sound Power Level to a "Base Reference Analog Sound Power Level". Like the two preceding equations, this one also provides **Sound Power Levels** in units of decibels, since the magnitudes of these values can also vary over a number of orders of magnitude.

$$L_P = 10\log\left[\frac{P}{P_0}\right] = 10\log\left[\frac{P}{10^{-12}}\right] = 10\log P + 120$$

Where: L_P =   the **Sound Power Level**, measured in decibels (dBs);

P = the measured Analog Sound Power Level, measured in units of watts (wts); and

P_0 = the "Base Reference Analog Sound Power Level," which has been set to have a value of 10^{-12} wts.

Sound Pressure Levels of Noise Sources in a Free Field

Equation 6.5

The following expression, Equation 6.5, identifies and relates the specific factors that must be accounted for when one determines the **Effective Sound Pressure Level** of any noise source in a "Free Field". For reference, a "Free Field" is any region (within which the noise source is located) that can be characterized as being free or void of any and all objects other than the noise source itself. Such a region permits the unhindered propagation of sound from the source in ALL directions. Because noise sources in the real world are seldom, if ever, in a true "Free Field", this expression has, as its final additive factor, a logarithmic term that effectively adjusts the resultant **Effective Sound Pressure Level** for any asymmetry that may exist in a real world "Non-Free Field" situation. Such factors effectively modify the "Free Fieldness" of the region where the noise source is located. As an example, a bell mounted on a wall would not be able to radiate sound in the direction of the wall; rather, it would effectively radiate sound only into a single spatial hemisphere. The factor that is used to achieve this result modification is called the Directionality Factor, and is defined below.

$$L_{P\text{-Effective}} = L_{P\text{-Source}} - 20\log r - 0.5 + 10\log Q$$

Where: $L_{P\text{-Effective}}$ = the **Effective Sound Pressure Level**, evaluated at a point that is "r" feet distant from the noise source itself, measured in decibels (dBs);

$L_{P\text{-Source}}$ = the source Sound Pressure Level, also measured in decibels (dBs);

r = the distance from the point where the **Effective Sound Pressure Level** is to be measured, to the noise source, measured in feet (ft); and

Q = the Directionality Factor, a dimensionless parameter — as defined and valued:

Q	=	**1**	for "spherical omnidirectional" radiating sources;
Q	=	**2**	for "single hemisphere" radiating sources;
Q	=	**4**	for "single quadrant" radiating sources;
Q	=	**8**	for "single octant" radiating sources

Addition of Sound Pressure Levels from Several Independent Sources
Equation 6.6

The following expression, Equation 6.6, is one of the most frequently employed relationships in all of acoustical engineering. It provides the basic methodology for determining the cumulative effect of several noise sources (each producing noise at an identifiable Sound Pressure Level) on an observer. This relationship provides for the determination of the **Effective Sound Pressure Level** that would be experienced by an observer from the several noise sources.

For this determination, we assume the perspective of an observer whose relative location — among two or more noise sources — causes him or her to experience an overall noise exposure — i.e., an **Effective Sound Pressure Level** — that will be: (1) obviously greater than would have been for a situation involving only a single noise source, but (2) certainly not simply the sum of the several sound pressure levels. As a descriptive example, consider the starter at a drag race. Assume this person is standing midway between two mufflerless dragsters, each producing sound at 130 dBA. Clearly this person would be exposed to a sound pressure level greater than 130 dBA, but not simply the sum value of 260 dBA. **Equation 6.6** provides the solution to the addition of sound pressure levels from several different separate and independent sources.

$$L_{total} = 10\log\left[\sum_{i=1}^{n} 10^{[L_i/10]}\right]$$

or

$$L_{total} = 10\log\left[10^{[L_1/10]} + 10^{[L_2/10]} + \ldots + 10^{[L_n/10]}\right]$$

Where: L_{total} =   the total **Effective Sound Pressure Level** resulting from the "n" different noise sources, measured in decibels (dBs); and

L_i = the Sound Pressure Level of the ith of "n" different noise sources, also measured in decibels (dBs).

Calculations Involving Sound Pressure Level "Doses"
Equation 6.7

The following expression, Equation 6.7, provides for the determination of the **Maximum Time Period** any worker may be safely exposed to some specifically quantified and/or **Equivalent Sound Pressure Level** from any number of noise sources.

$$T_{max} = \frac{8}{2^{[(L - 90)/5]}}$$

Where: T_{max} = the **Maximum Time Period** — at any Equivalent Sound Pressure Level, L — to which a worker may be exposed during a normal eight-hour workday, measured in some convenient time unit, usually hours; and

L = the Sound Pressure Level (or Equivalent Sound Pressure Level) being evaluated for this situation, measured in decibels (dBs).

Equation 6.8

The second relationship involving "doses" is Equation 6.8, which provides the basis for calculating the **Effective Daily Dose** that an individual would have experienced as a result of his or her having been exposed to several different well-quantified sound pressure levels, each of which occurred for some specific Time Period or Time Interval.

$$D = \sum_{i=1}^{n} \frac{C_i}{T_{max_i}} = \frac{C_1}{T_{max_1}} + \frac{C_2}{T_{max_2}} + \ldots + \frac{C_n}{T_{max_n}}$$

Where: D = the **Effective Daily Dose** (of noise) to which an individual who has been exposed to a series of "n" different sound pressure levels, with each of these exposures lasting for a known Time Period or Time Interval, C_i; the **Effective Daily Dose**, **D**, is a dimensionless decimal number;

C_i = the overall ith Time Interval or Time Period during which the individual being considered was exposed to the ith sound pressure level; these time intervals will always have to be measured in some consistent unit of time, usually in hours; and

T_{max_i} = the Maximum Time Period that would be permitted for the ith specific sound pressure level to which an individual could be exposed; as defined by Equation 6.7.

Note For any value of **Effective Daily Dose, D** ≤ 1.00, the individual who has experienced this dose will have accumulated neither an excessive nor a harmful amount of noise. On the other hand, if this parameter assumes a value greater than 1.00, then the exposure would have to be classified as potentially harmful.

Equation 6.9

The following expression, Equation 6.9, provides the relationship for determining the **Equivalent Sound Pressure Level** that corresponds to any identified **Daily Dose**.

$$L_{equivalent} = 90 + 16.61 \log D$$

Where: $L_{equivalent}$ = the **Equivalent Sound Pressure Level** that corresponds to any Daily Dose, with this parameter measured in decibels (dBs); and

D = the Daily Dose, as defined by Equation **6.8**.

Definitions of the Three Common Frequency Weighting Scales

Equation 6.10

The following tabular listing serves as the defining descriptor for each of the three commonly used Frequency Weightings, as these weightings are applied to the measurement of any sound or noise. These weightings, which are applied to the specific Full or Unitary Octave Bands that make up the sound that is being monitored, are designed to provide a measured result that approximates the response of the human ear to sounds or noises of various intensities. Specifically, the **A-Weighting Scale** is thought to provide a result that approximates the response of the human ear to low intensity sounds or noises. The **B-Weighting Scale** provides a human ear based response to sounds or noises having a moderate or medium intensity. Finally, the **C-Weighting Scale** is thought to provide a similar result when applied to high intensity sounds or noises. The **A-Weighting Scale** is, by far, the most widely used of these three; the other two are now only rarely used.

All three Frequency Weighting Sales are in the form of "additions to" or "deductions from" the Full or Unitary Octave Bands that make up the sound that is being monitored. Whenever the sound level meter being used to monitor some sound or noise has been set up to provide a result to which one of these Frequency Weightings has been applied, the resultant units of the measurement must have — as applicable — an "A", a "B", or a "C" appended to it — i.e., dBA for an **A-Weighting Scale** measurement, dBB for the **B-Weighting Scale**, and/or dBC for the **C-Weighting Scale**.

The following table shows the additions or deductions that must be applied to the various Octave Bands in order to make the required Frequency Weighting adjustments.

Full Octave	*(Deductions) or Increments, in decibels*		
Band, in Hertz	A-Scale	B-Scale	C-Scale
31 Hz	(39)	(17)	(3)
63 Hz	(26)	(9)	(1)
125 Hz	(16)	(4)	0
250 Hz	(9)	(1)	0
500 Hz	(3)	0	0
1,000 Hz	0	0	0
2,000 Hz	1	0	0
4,000 Hz	1	(1)	(1)
8,000 Hz	(1)	(3)	(3)

Various Octave Band Relationships
Equations 11 and 12

The following two relationships, namely, Equations 6.11 and 6.12, identify the specific relationships that apply to any Full or Unitary Octave Band, as specified by ANSI S1.11-1966 (R1975). *Note:* The nine Full or Unitary Octave Bands listed in the table above — as part of the Definitions of the three Frequency Weighting Scales — are the commonly accepted Full or Unitary Octave Bands. For the overall set of Full or Unitary Octave Bands, the Center Frequency of the "Middle" Octave Band is 1,000 Hz. Each member of this set of nine Full or Unitary Octave Bands will have the following characteristics:

1. Each band will be one full octave in total Bandwidth — i.e., the Band's Lower Band-Edge Frequency will always be half of its Upper Band-Edge Frequency.
2. The Center Frequency of each band will be the "Geometric Mean" of its Lower and its Upper Band-Edge Frequencies.
3. Each band will have a Center Frequency that will be one half of the Center Frequency of the next higher Octave Band, and twice the Center Frequency of the next lower one.

The first two of these three overall relationships can be expressed quantitatively and are shown below.

Equation 6.11

$$f_{upper-1/1} = 2\left(f_{lower-1/1}\right)$$

Equation 6.12

$$f_{center-1/1} = \sqrt{\left(f_{upper-1/1}\right)\left(f_{lower-1/1}\right)} = \text{the "Geometric Mean"}$$

Where: $f_{upper-1/1}$ =   the Upper Band-Edge Frequency for the specific Full or Unitary Octave Band being considered;

        $f_{lower-1/1}$ =   the Lower Band-Edge Frequency for the specific Full or Unitary Octave Band being considered; and

        $f_{center-1/1}$ =   the Center Frequency for the specific Full or Unitary Octave Band being considered.

Equations 6.13 and 6.14

The following two relationships, Equations 6.13 and 6.14, identify the specifics of a Standard Half Octave Band, also as specified by ANSI S1.11-1966 (R1975). For the overall set of Half Octave Bands, the Center Frequency of the "Middle" Octave Band is, like its Full or Unitary Octave Band counterpart, at 1,000 Hz. For the overall set of Half Octave Bands, the following set of characteristics always applies:

1. Each Half Octave Band will be $1/\sqrt{2}$ Octaves in total Bandwidth, i.e., the Lower Band-Edge Frequency of each Half Octave Band will always be $1/\sqrt{2}$ of its Upper Band-Edge Frequency.
2. The Center Frequency of each Half Octave Band will be the "Geometric Mean" of its Lower and its Upper Band-Edge Frequencies.

3. Each Half Octave Band will have a Center Frequency that will be $1/\sqrt{2}$ of the Center Frequency of the next higher Half Octave Band, and $\sqrt{2}$ times the Center Frequency of the next lower one.

Again the first two of these three overall relationships can be expressed quantitatively and are shown below.

Equation 6.13

$$f_{upper-1/2} = \left(\sqrt{2}\right)\left(f_{lower-1/2}\right)$$

Equation 6.14

$$f_{center-1/2} = \sqrt{\left(f_{upper-1/2}\right)\left(f_{lower-1/2}\right)} = \text{the "Geometric mean"}$$

Where: $\mathbf{f_{upper-1/2}}$ =   the Upper Band-Edge Frequency for the specific Half Octave Band being considered;

$\mathbf{f_{lower-1/2}}$ =   the Lower Band-Edge Frequency for the specific Half Octave Band being considered; and

$\mathbf{f_{center-1/2}}$ =   the Center Frequency for the specific Half Octave Band being considered.

Equations 6.15 and 6.16

The following two relationships, **Equations 6.15** and **6.16**, identify the specifics of a "$1/n^{th}$" Octave Band, also as specified by ANSI S1.11-1966 (R1975). For the overall set of $1/n^{th}$ Octave Bands, the Center Frequency of the "Middle" Band is, like its other counterparts, at 1,000 Hz. For the overall set of $1/n^{th}$ Octave Bands, the following set of characteristics always applies:

1. Each $1/n^{th}$ Octave Band will be $1/\sqrt[n]{2}$ Octaves in total Bandwidth, i.e., the Lower Band-Edge Frequency of each $1/n^{th}$ Octave Band will always be $1/\sqrt[n]{2}$ of its Upper Band-Edge Frequency.
2. The Center Frequency of each $1/n^{th}$ Octave Band will be the "Geometric Mean" of its Lower and its Upper Band-Edge Frequencies.
3. Each $1/n^{th}$ Octave Band will have a Center Frequency that will be $1/\sqrt[n]{2}$ of the Center Frequency of the next higher Half Octave Band, and $\sqrt[n]{2}$ times the Center Frequency of the next lower one.

Again the first two of these three overall relationships can be expressed quantitatively and are shown next.

Equation 6.15

$$f_{upper-1/n} = \left(\sqrt[n]{2}\right)\left(f_{lower-1/n}\right)$$

Equation 6.16

$$f_{center-1/n} = \sqrt[n]{\left(f_{upper-1/n}\right)\left(f_{lower-1/n}\right)} = \text{the "Geometric Mean"}$$

Where: $f_{upper-1/n}$ =   the Upper Band-Edge Frequency for the specific $1/n^{th}$ Octave Band being considered;

          $f_{lower-1/n}$ =   the Lower Band-Edge Frequency for the specific $1/n^{th}$ Octave Band being considered; and

          $f_{center-1/n}$ =   the Center Frequency for the specific $1/n^{th}$ Octave Band being considered.

Chapter 7
Ionizing and Nonionizing Radiation

Interest in this area of potential human hazard stems, in part, from the magnitude of harm or damage that an individual who is exposed can experience. It is widely known that the risks associated with exposures to ionizing radiation are significantly greater than comparable exposures to nonionizing radiation. This fact notwithstanding, it is steadily becoming more widely accepted that nonionizing radiation exposures also involve risks to which one must pay close attention. This chapter will focus on the fundamental characteristics of the various types of ionizing and nonionizing radiation, as well as on the factors, parameters, and relationships whose application will permit accurate assessments of the hazard that might result from exposures to any of these physical agents.

Relevant Definitions
Electromagnetic Radiation

Electromagnetic Radiation refers to the entire spectrum of photonic radiation, from wavelengths of less than 10^{-5} Å (10^{-15} meters) to those greater than 10^8 meters — a dynamic wavelength range of more than 22+ decimal orders of magnitude. It includes all of the segments that make up the two principal sub-categories of this overall spectrum, which are the "Ionizing" and the "Nonionizing" radiation sectors. Photons having wavelengths shorter than 0.4 μ (400 nm or 4,000 Å) fall under the category of Ionizing Radiation; those with longer wavelengths will all be in the Nonionizing group. In addition, the overall Nonionizing Radiation sector is further divided into the following three sub-sectors:

Optical Radiation Band*	0.1 μ to 2,000 μ, *or*
	0.0001 to 2.0 mm
Radio Frequency/Microwave Band	2.0 mm to 10,000,000 mm, *or*
	0.002 to 10,000 m
Sub-Radio Frequency Band	10,000 m to 10,000,000+ m, *or*
	10 km to 10,000+ km

It must be noted that the entirety of the ultraviolet sector (0.1 μ to 0.4 μ wavelengths) is listed as a member of the Optical Radiation Band, and appears, therefore, to be a Nonionizing type of radiation. This is not true. UV radiation is indeed ionizing; it is just categorized incorrectly insofar as its group membership among all the sectors of **Electromagnetic Radiation.*

Although the discussion thus far has focused on the wavelengths of these various bands, this subject also has been approached from the perspective of the frequencies involved. Not surprisingly, the dynamic range of the frequencies that characterize the entire **Electromagnetic Radiation** spectrum also covers 22+ decimal orders of magnitude — ranging from 30,000 exahertz or 3×10^{22} hertz (for the most energetic cosmic rays) to approximately 1 or 2 hertz (for the longest wavelength ELF photons). The energy of any photon in this overall spectrum will be directly proportional to its wavelength — i.e., photons with the highest frequency will be the most energetic.

The most common **Electromagnetic Radiation** bands are shown in a tabular listing on the following page. Table 1 utilizes increasing wavelengths, or λs, as the basis for identifying each spectral band.

Ionizing Radiation

Ionizing Radiation is any photonic (or particulate) radiation — either produced naturally or by some man-made process — that is capable of producing or generating ions. Only the shortest wavelength (highest energy) segments of the overall electromagnetic spectrum are capable of interacting with other forms of matter to produce ions. Included in this grouping are most of the ultraviolet band (even though this band is catalogued in the Nonionizing sub-category of Optical Radiation), as well as every other band of photonic radiation having wavelengths shorter than those in the UV band.

Ionizations produced by this class of electromagnetic radiation can occur either "directly" or "indirectly". "Directly" ionizing radiation includes:

1. electrically charged particles (i.e., electrons, positrons, protons, α-particles, etc.), and
2. photons/particles of sufficiently great kinetic energy that produce ionizations by colliding with atoms and/or molecules present in the matter.

In contrast, "indirectly" ionizing particles are always uncharged (i.e., neutrons, photons, etc.). They produce ionizations indirectly, either by:

1. liberating one or more "directly" ionizing particles from matter with which these particles have interacted or are penetrating, or
2. initiating some sort of nuclear transition or transformation (i.e., radioactive decay, fission, etc.) as a result of their interaction with the matter through which these particles are passing.

Protection from the adverse effects of exposure to various types of **Ionizing Radiation** is an issue of considerable concern to the occupational safety and health professional. Certain types of this class of radiation can be very penetrating (i.e., γ-rays, X-rays, and neutrons); that is to say these particles will typically require very substantial shielding in order to ensure the safety of workers who might otherwise become exposed. In contrast to these very penetrating forms of **Ionizing Radiation**, α- and β-particles are far less penetrating, and therefore require much less shielding.

Table 1.   Electromagnetic Radiation Bands

	Photon Wavelength, λ	
Spectral Band	*Band Min. λ*	*Band Max. λ*
Ionizing Radiation		
Cosmic Rays	<0.00005 Å	0.005 Å
γ-rays	0.005 Å	0.8 Å
X-rays — hard	0.8 Å	5.0 Å
X-rays — soft	5.0 Å	80 Å
	0.5 nm	8.0 nm
Nonionizing Radiation		
Optical Radiation Bands		
Ultraviolet — UV-C	8.0 nm	250 nm
	0.008 μ	0.25 μ
Ultraviolet — UV-B	250 nm	320 nm
	0.25 μ	0.32 μ
Ultraviolet — UV-A	320 nm	400 nm
	0.32 μ	0.4 μ
Visible Light	0.4 μ	0.75 μ
Infrared — Near or IR-A	0.75 μ	2.0 μ
Infrared — Mid or IR-B	2.0 μ	20 μ
Infrared — Far or IR-C	20 μ	2,000 μ
	0.02 mm	2 mm
Radio Frequency/Microwave Bands		
Extremely High Frequency (EHF) *Microwave* Band	1 mm	10 mm
Super High Frequency (SHF) *Microwave* Band	10 mm	100 mm
Ultra High Frequency (UHF) *Microwave* Band	100 mm	1,000 mm
	0.1 m	1 m
Very High Frequency (VHF) *Radio Frequency* Band	1 m	10 m
High Frequency (HF) *Radio Frequency* Band	10 m	100 m
Medium Frequency (MF) *Radio Frequency* Band	100 m	1,000 m
	0.1 km	1 km
Low Frequency (LF) Band	1 km	10 km
Sub-Radio Frequency Bands		
Very Low Frequency (VLF) Band	10 km	100 km
Ultra Low Frequency (ULF) Band	100 km	1,000 km
	0.1 Mm	1 Mm
Super Low Frequency (SLF) Band	1 Mm	10 Mm
Extremely Low Frequency (ELF) *Power Freqency* Band	10 Mm	>100 Mm

Categories of Ionizing Radiation
Cosmic Radiation

Cosmic Radiation (cosmic rays) makes up the most energetic — therefore, potentially the most hazardous — form of Ionizing Radiation. **Cosmic Radiation** consists primarily of high speed, very high energy protons (protons with velocities approaching the speed of light) — many or even most with energies in the billions or even trillions of electron volts. These particles originate at various locations throughout space, eventually arriving on the earth after traveling great distances from their "birthplaces". Cataclysmic events, or in fact any event in the universe that liberates large amounts of energy (i.e., supernovae, quasars, etc.), will be sources of **Cosmic Radiation**. It is fortunate that the rate of arrival of cosmic rays on Earth is very low; thus the overall, generalized risk to humans of damage from cosmic rays also is relatively low.

Nuclear Radiation

Nuclear Radiation is, by definition, terrestrial radiation that originates in, and emanates from, the nuclei of atoms. From one perspective then, this category of radiation probably should not be classified as a subset of electromagnetic radiation, since the latter is made up of photons of pure energy, whereas **Nuclear Radiation** can be either energetic photons or particles possessing mass (i.e., electrons, neutrons, helium nuclei, etc.). It is clear, however, that this class of "radiation" does belong in the overall category of Ionizing Radiation; thus it will be discussed here. In addition, according to Albert Einstein's Relativity Theory, energy and mass are equivalent — simplistically expressed as $E = mc^2$ — this fact further solidifies the inclusion of **Nuclear Radiation** in this area.

Nuclear events such as radioactive decay, fission, etc. all serve as sources for **Nuclear Radiation**. Gamma rays, X-rays, alpha particles, beta particles, protons, neutrons, etc., as stated on the previous page, can all be forms of **Nuclear Radiation**. Cosmic rays should also be included as a subset in this overall category, since they clearly originate from a wide variety of nuclear sources, reactions, and/or disintegrations; however, since they are extra-terrestrial in origin, they are not thought of as **Nuclear Radiation**. Although of interest to the average occupational safety and health professional, control and monitoring of this class of ionizing radiation usually falls into the domain of the Health Physicist.

Gamma Radiation

Gamma Radiation — Gamma rays (γ-rays) — consists of very high energy photons that have originated, most probably, from one of the following four sources:

1. nuclear fission (i.e., the explosion of a simple "atomic bomb", or the reactions that occur in a power generating nuclear reactor),

2. nuclear fusion (i.e., the reactions that occur during the explosion of a fusion based "hydrogen bomb", or the energy producing mechanisms of a star, or the operation of one of the various experimental fusion reaction pilot plants, the goal of which is the production of a self-sustaining nuclear fusion-based source of power),

3. the operation of various fundamental particle accelerators (i.e., electron linear accelerators, heavy ion linear accelerators, proton synchrotrons, etc.), or

4. the decay of a radionuclide.

While there are clearly four well-defined source categories for **Gamma Radiation**, the one upon which we will focus will be the decay of a radio-active nucleus. Most of the radioactive decays that produce γ-Rays also produce other forms of ionizing radiation (β⁻-particles, principally); however, the practical uses of these radionuclides rest mainly on their γ-Ray emissions. The most common application of this class of isotope is in the medical area. Included among the radionuclides that have applications in this area are: $_{53}^{125}$I and $_{53}^{131}$I (both used in thyroid therapy), and $_{27}^{60}$Co (often used as a source of high energy γ-Rays in radiation treatments for certain cancers).

Gamma rays are uncharged, highly energetic photons possessing usually 100+ times the energy, and less than 1% of the wavelength, of a typical X-ray. They are very penetrating, typically requiring a substantial thickness of some shielding material (i.e., lead, steel reinforced concrete, etc.).

Alpha Radiation

Alpha Radiation — Alpha Rays (α-Rays, α-particles) — consists solely of the completely ionized nuclei of helium atoms, generally in a high energy condition. As such, α-Rays are particulate and not simply pure energy; thus they should not be considered to be electromagnetic radiation — see the discussion under the previous topic of Nuclear Radiation.

These nuclei consist of two protons and two neutrons each, and as such, they are among the heaviest particles that one ever encounters in the nuclear radiation field. The mass of an α-particle is 4.00 atomic mass units, and its charge is +2 (twice the charge of the electron, but positive — the basic charge of an electron is -1.6×10^{-19} coulombs). The radioactive decay of many of the heaviest isotopes in the periodic table frequently involves the emission of α-particles. Among the nuclides included in this grouping are $_{92}^{238}$U, $_{88}^{226}$Ra, and $_{86}^{222}$Rn.

Considered as a member of the nuclear radiation family, the α-particle is the least penetrating. Typically, **Alpha Radiation** can be stopped by a sheet of paper; thus, shielding individuals from exposures to α-particles is relatively easy. The principal danger to humans arising from exposures to α-particles occurs when some alpha active radionuclide is ingested and becomes situated in some vital organ in the body where its lack of penetrating power no longer is a factor.

Beta Radiation

Beta Radiation constitutes a second major class of directly ionizing charged particles; and again because of this fact, this class or radiation should not be considered to be a subset of electromagnetic radiation.

There are two different β-particles — the more common negatively charged one, the β^- (the electron), and its positive cousin, the β^+ (the positron). **Beta Radiation** most commonly arises from the radioactive decay of an unstable isotope. A radioisotope that decays by emitting β-particles is classified as being beta active. Among the most common beta active (all β^- active) radionuclides are $_1^3H$ (tritium), $_6^{14}C$, and $_{38}^{90}Sr$.

Most **Beta Radiation** is of the β^- category; however, there are radionuclides whose decay involves the emission of β^+ particles. β^+ emissions inevitably end up falling into the Electron Capture (EC) type of radioactive decay simply because the emitted positron — as the antimatter counterpart of the normal electron, or β^- particle — annihilates immediately upon encountering its antiparticle, a normal electron. Radionuclides that are β^+ active include: $_{11}^{22}Na$ and $_9^{18}F$.

Although more penetrating than an α-particle, the β-particle is still not a very penetrating form of nuclear radiation. β-particles can generally be stopped by very thin layers of any material of high mass density (i.e., 0.2 mm of lead), or by relatively thicker layers of more common, but less dense materials (i.e., a 1-inch thickness of wood). As is the case with α-particles, β-particles are most dangerous when an ingested beta active source becomes situated in some susceptible organ or other location within the body.

Neutron Radiation

Although there are no naturally occurring neutron sources, this particle still constitutes an important form of nuclear radiation; and again since the neutron is a massive particle, it should not simply be considered to be a form of electromagnetic radiation. As was the case with both α- and β-particles, neutrons can generate ions as they interact with matter; thus they definitely are a subset of the overall class of ionizing radiation. The most important source of **Neutron Radiation** is the nuclear reactor (commercial, research, and/or military). The characteristic, self-sustaining chain reaction of an operating nuclear reactor, by definition, generates a steady supply of neutrons. Particle accelerators also can be a source of **Neutron Radiation**.

Protecting personnel from exposures arising from **Neutron Radiation** is one of the most difficult problems in the overall area of radiation protection. Neutrons can produce considerable damage in exposed individuals. Unlike their electrically charged counterparts (α- and β-particles), uncharged neutrons are not capable, either directly or indirectly, of producing ionizations. Additionally, neutrons do not behave like high energy photons (γ-Rays and/or X-rays) as they interact with matter. These relatively massive uncharged particles simply pass through matter without producing anything until they collide with one of the nuclei that are resident there. These collisions accomplish two things simultaneously:

1. they reduce the energy of the neutron, and

2. they "blast" the target nucleus, usually damaging it in some very significant manner — i.e., they mutate this target nucleus into an isotope of the same element that has a higher atomic weight, one that will likely be radioactive. Alternatively, if neutrons are passing through some fissile material, they can initiate and/or maintain a fission chain reaction, etc.

Shielding against **Neutron Radiation** always involves processes that reduce the energy or the momentum of the penetrating neutron to a point where its collisions are no longer capable of producing damage. High energy neutrons are most effectively attenuated (i.e., reduced in energy or momentum) when they collide with an object having approximately their same mass. Such collisions reduce the neutron's energy in a very efficient manner. Because of this fact, one of the most effective shielding media for neutrons is water, which obviously contains large numbers of hydrogen nuclei, or protons which have virtually the same mass as the neutron.

X-radiation

X-radiation — X-rays — consists of high energy photons that, by definition, are man-made. The most obvious source of X-radiation is the X-ray machine, which produces these energetic photons as a result of the bombardment of certain heavy metals — i.e., tungsten, iron, etc. — with high energy electrons. X-rays are produced in one or the other of the two separate and distinct processes described below:

1. the acceleration (actually, negative acceleration or "deceleration") of a fast moving, high energy, negatively charged electron as it passes closely by the positively charged nucleus of one of the atoms of the metal matrix that is being bombarded (energetic X-ray photons produced by this mechanism are known as "Bremsstrahlung X-rays", and their energy ranges will vary according to the magnitude of the deceleration experienced by the bombarding electron); and

2. the de-excitation of an ionized atom — an atom that was ionized by a bombarding, high energy electron, which produced the ionization by "blasting" out one of the target atom's own inner shell electrons — the de-excitation occurs when one of the target atom's remaining outer shell electrons "falls" into (transitions into) the vacant inner shell position, thereby producing an X-ray with an energy precisely equal to the energy difference between the beginning and ending states of the target atom (energetic X-ray photons produced in this manner are known as "Characteristic X-rays" because their energies are always precisely known).

The principal uses of **X-Radiation** are in the areas of medical and industrial radiological diagnostics. The majority of the overall public's exposure to ionizing radiation occurs as a result of exposure to X-rays.

Like their γ-ray counterparts, X-rays are uncharged, energetic photons with substantial penetrating power, typically requiring a substantial thickness of some shielding material (i.e., lead, iron, steel reinforced concrete, etc.) to protect individuals who might otherwise be exposed.

Ultraviolet Radiation

Photons in the **Ultraviolet Radiation** or UV spectral band have the least energy that is still capable of producing ionizations. As stated earlier, all of the UV band has been classified as being a member of the *Optical Radiation Band*, which — by definition — is Nonionizing. This is erroneous, since UV is indeed capable of producing ionizations in exposed matter. Photoionization detection, as a basic analytical tool, relies on the ability of certain wavelengths of UV radiation to generate ions in certain gaseous components.

"Black Light" is a form of **Ultraviolet Radiation**. In the industrial area, UV radiation is produced by plasma torches, arc welding equipment, and mercury discharge lamps. The most prominent source of UV is the Sun.

Ultraviolet Radiation has been further classified into three subcategories by the *Commission Internationale d'Eclairage* (CIE). These CIE names are: UV-A, UV-B, and UV-C. The wavelengths associated with each of these "CIE Bands" are shown in Table 1.

The UV-A band is the least dangerous of these three, but it has been shown to produce cataracts in exposed eyes. UV-B and UV-C are the bands responsible for producing injuries such as photokeratitis (i.e., welder's flash, etc.), and erythema (i.e., sunburn, etc.). A variety of protective measures are available to individuals who may become exposed to potentially harmful UV radiation. Included among these methods are glasses or skin ointments designed to block harmful UV-B and/or UV-C photons.

Categories of Nonionizing Radiation
Visible Light

Visible Light is that portion of the overall electromagnetic spectrum to which our eyes are sensitive. This narrow spectral segment is the central member of the *Optical Radiation Band*. The hazards associated with **Visible Light** depend upon a combination of the energy of the source and the duration of the exposure. Certain combinations of these factors can pose very significant hazards (i.e., night and color vision impairments). In cases of extreme exposure, blindness can result. As an example, it would be very harmful to an individual's vision for that individual to stare, even for a very brief time period, at the sun without using some sort of eye protection. In the same vein, individuals who must work with visible light lasers must always wear protective glasses — i.e., glasses with appropriate optical density characteristics.

For reference, the retina, which is that part of the eye that is responsible for our visual capabilities, can receive the entire spectrum of visible light as well as the near infrared — which will be discussed under the next definition. It is the exposure to these bands that can result in vision problems for unprotected individuals.

Infrared Radiation

Infrared Radiation, or IR, is the longest wavelength sector of the overall *Optical Radiation Band*. The IR spectral band, like its UV relative, is usually thought of as being divided into three sub-segments, the near, the mid, and the far. These three sub-bands have also been designated by the *Commission Internationale d'Eclairage* (CIE), respectively, as IR-A, IR-B, and IR-C. The referenced non-CIE names, "near", "mid", and "far", refer to the relative position of the specific IR band with respect to visible light — i.e., the near IR band has wavelengths that are immediately adjacent to the longest visible light wavelengths, while the far IR photons, which have the greatest infrared wavelengths, are most distant from the visible band. In general, we experience **Infrared Radiation** as radiant heat.

As stated earlier in the discussion for visible light, the anterior portions of the eye (i.e., the lens, the vitreous humor, the cornea, etc.) are all largely opaque to the mid and the far IR; only the photons of the near IR can penetrate all the way to the retina. Near IR photons are, therefore, responsible for producing retinal burns. Mid and far IR band photons, for which the anterior portions of the eye are relatively opaque, will typically be absorbed in these tissues and are, therefore, responsible for injuries such as corneal burns.

Microwave Radiation

General agreement holds that **Microwave Radiation** involves the EHF, SHF, and UHF Bands, plus the shortest wavelength portions of the VHF Band — basically, the shortest wavelength half of the *Radio Frequency/Microwave Band* sub-group. All the members of this group have relatively short wavelengths — the maximum λ is in the range of three meters.

Virtually all the adverse physiological effects or injuries that accrue to individuals who have been exposed to harmful levels of **Microwave Radiation** can be understood from the perspective of the "radiation" rather than the "electric and/or magnetic field" characteristics of these physical agents. (See the discussion of the differences between these two characteristic categories, as well as the associated concepts of the "Near Field" and the "Far Field", later in this chapter under *Radiation Characteristics vs. Field Characteristics*). Physiological injuries to exposed individuals, to the extent that they occur at all, are simply the result of the absorption — within the body of the individual who has been exposed to the **Microwave Radiation** — of a sufficiently large amount of energy to produce significant heating in the exposed organs or body parts. The long-term health effects of exposures that do not produce any measurable heating (i.e., increases in the temperature of some organ or body part) are unknown at this time.

Some of the uses/applications that make up each of the previously identified **Microwave Radiation** bands are listed in the following table:

Band	Wavelength	Frequency	Use or Application
EHF	1–10 mm	300–30 GHz	Satellite Navigational Aids and Communications, Police 35 GHz *K Band* Radar, Microwave Relay Stations, Radar: *K (partial), L and M Bands* (military fire control), High Frequency Radio, etc.
SHF	10–100 mm	30–3 GHz	Police 10 and 24 GHz *J and K Band* Radars, Satellite Communications, Radar: *F, G, H, I, J, and K (partial) Bands* (surveillance, and marine applications), etc.
UHF	0.1–1.0 m	3,000–300 MHz	UHF Television (Channels 14 to 84), certain CB Radios, Cellular Phones, Microwave Ovens, Radar: *B (partial), D, and E Bands* (acquisition and tracking, plus air traffic control), Taxicab Communications, Spectroscopic Instruments, some Short-wave Radios, etc.
VHF	1.0–3.0 m	300–100 MHz	Higher Broadcast Frequency Standard Television (174 to 216 MHz: Channels 7 to 13), Radar *B Band*, Higher Frequency FM Radio (100+ MHz), walkie-talkies, certain CB Radios, Cellular Telephones, etc.

Radio Frequency Radiation

Radio Frequency Radiation makes up the balance of the *Radio Frequency/Microwave Band* sub-group. The specific segments involved are the longest wavelength half of the VHF Band, plus all of the HF, MF, and LF Bands. In general, all of the wavelengths involved in this sub-group are considered to be long to very long, with the shortest λ being 3+ meters and the longest, approximately 10 km, or just less than 6.25 miles.

The adverse physiological effects or injuries, if any, that result from exposures to **Radio Frequency Radiation** can be understood from the perspective of the "electric and/or magnetic field", rather than the "radiation" characteristics of these particular physical agents (again, see discussion of the differences in these two characteristic categories, as well as the associated concepts of the "Near Field" and the "Far Field", under *Radiation Characteristics vs. Field Characteristics*). Injuries to exposed indi-viduals, to the extent that they have been documented at all, also are the result of the absorption by some specific organ or body part of a sufficiently large amount of energy to produce highly localized heating. As was the case with Microwave Radiation exposures, the long-term health effects of exposure events that do not produce any measurable heating are unknown at this time.

Some of the uses/applications that make up each of the previously identified **Radio Frequency Radiation** bands are listed in the following table:

Band	Wavelength	Frequency	Use or Application
VHF	3.0–10.0 m	100–30 MHz	Lower Frequency Broadcast Standard Television (54 to 72, and 76 to 88 MHz: Channels 2 to 6), Lower Frequency FM Radio (88 to 100 MHz), Dielectric Heaters, Diathermy Machines, certain CB Radios, certain Cellular Telephones, etc.
HF	10–100 m	30–3 MHz	Plasma Processors, Dielectric Heaters, various types of Welding, some Short-wave Radios, Heat Sealers, etc.
MF	0.1–1.0 km	3,000–300 kHz	Plasma Processors, AM Radio, various types of Welding, some Short-wave Radios, etc.
LF	1–10 km	300–30 kHz	Cathode Ray Tubes or Video Display Terminals

Sub-Radio Frequency Radiation

The final portion of the overall electromagnetic spectrum is comprised of its longest wavelength members. **Sub-Radio Frequency Radiation** makes up its own "named" category, namely, the *Sub-Radio Frequency Band*, as the final sub-group of the overall category of Nonionizing Radiation.

At the time that this paragraph is being written, there is little agreement as to the adverse physiological effects that might result from exposures to **Sub-Radio Frequency Radiation**. Again, and to the extent that human hazards do exist for this class of physical agent, these hazards can be best understood from the perspective of the "electric and/or magnetic field", rather than the "radiation" characteristics of **Sub-Radio Frequency Radiation** (See the discussion of the differences between these two characteristic categories, as well as the associated concepts of the "Near Field" and the "Far Field" under *Radiation Characteristics vs. Field Characteristics*.

Primary concern in this area seems generally to be related to the strength of either or both the electric and the magnetic fields that are produced by sources of this class of radiation. The American Conference of Government Industrial Hygienists (ACGIH) has published the following expressions that can be used to calculate the appropriate eight-hour TLV-TWA — each as a function of the frequency, **f**, of the **Sub-Radio Frequency Radiation** source being considered. The relationship for electric fields provides a field strength TLV expressed in volts/meter (V/m); while the relationship for magnetic fields produces a magnetic flux density TLV in milliteslas (mT).

$$\underline{\text{Electric Fields} \qquad\qquad \text{Magnetic Fields}}$$

$$E_{TLV} = \frac{2.5 \times 10^6}{f} \qquad\qquad B_{TLV} = \frac{60}{f}$$

Finally, one area where there does appear to be very considerable, well-founded concern about the hazards produced by **Sub-Radio Frequency Radiation** is in the area of the adverse impacts of the electric and magnetic fields produced by this class of source on the normal operation of cardiac pacemakers. An electric field of 2,500 volts/meter (2.5 kV/m) and/or a magnetic flux density of 1.0 gauss (1.0 G, which is equivalent to 0.1 milliteslas or 0.1 mT) each clearly has the potential for interrupting the normal operation of an exposed cardiac pacemaker, virtually all of which operate at roughly these same frequencies.

Some of the uses/applications that make up each of the previously identified **Sub-Radio Frequency Radiation** bands are listed in this table:

Band	Wavelength	Frequency	Use or Application
VLF	10–100 km	30–3 kHz	Cathode Ray Tubes or Video Display Terminals (video flyback frequencies), certain Cellular Telephones, Long-Range Navigational Aids (LORAN), etc.
ULF	0.1–1 Mm	3,000–300 Hz	Induction Heaters, etc.
SLF	1–10 Mm	300–30 Hz	Standard Electrical Power (60 Hz), Home Appliances, Underwater Submarine Communications, etc.
ELF	10–100 Mm	30–3 Hz	Underwater Submarine Communications, etc.

Radiation Characteristics vs. Field Characteristics

All of the previous discussions have been focused on the various categories and sub-categories of the electromagnetic spectrum (excluding, in general, the category of particulate nuclear radiation). It must be noted that every band of electromagnetic radiation — from the extremely high frequencies of Cosmic Rays (frequencies often greater than 3×10^{21} Hz or 3,000 EHz) to the very low end frequencies characteristic of normal electrical power in the United States (i.e., 60 Hz) — will consist of photons of **radiation** possessing both electric and magnetic **field** characteristics.

That is to say, we are dealing with **radiation** phenomena that possess **field** (electric and magnetic) characteristics. The reason for considering these two different aspects or factors is that measuring the "strength" or the "intensity" of any radiating source is a process in which only rarely will both the **radiation** and the **field** characteristics be easily quantifiable. The vast majority of measurements in this field will, of necessity, have to be made on only one or the other of these characteristics. It is the frequency and/or the wavelength being considered that determines whether the measurements will be made on the **radiation** or the **field** characteristics of the source involved.

When the source frequencies are relatively high — i.e., **f** > 100 MHz (with λ < 3 meters) — it will almost always be easier to treat and measure such sources as simple **radiation** sources. For these monitoring applications (with the exception of situations that involve lasers), it will be safe to assume that the required "strength" and/or "intensity" characteristics will behave like and can be treated as if they were **radiation** phenomena — i.e., they vary according to the inverse square law.

In contrast, when the source frequencies fall into the lower ranges — i.e., **f** ≤ 100 MHz (with λ ≥ 3 meters) — then it will be the **field** characteristics that these sources produce (electric and/or magnetic) that will be relatively easy to measure. While it is certainly true that these longer wavelength "photons" do behave according to the inverse square law — since they are, in fact, radiation — their relatively long wavelengths make it very difficult to measure them as radiation phenomena.

These measurement problems relate directly to the concepts of the Near and the Far Field. The Near Field is that region that is close to the source — i.e., no more than a very few wavelengths distant from it. The Far Field is the entire region that exists beyond the Near Field.

Field measurements (i.e., separate electric and/or magnetic field measurements) are usually relatively easy, so long as the measurements are completed in the Near Field. It is in this region where specific, separate, and distinct measurements of either of these two **fields** can be made. The electric **fields** that exist in the Near Field are produced by the voltage characteristics of the source, while the magnetic **fields** in this region result from the source's electrical current. Electric field strengths will typically be expressed in one of the following three sets of units: (1) volts/meter — v/m; (2) volts^2/meter^2 — v^2/m^2; or (3) milliwatts/cm^2 — mW/cm^2. Magnetic field intensities will typically be expressed in one of the following four sets of units: (1) amperes/meter — A/m; (2) milliamperes/meter — mA/m; (3) Amperes^2/meter^2 — A^2/m^2; or (4) milliwatts/cm — mW/cm.

Radiation measurements, in contrast, are typically always made in the Far Field. As an example, let us consider a 75,000 volt X-ray Machine — i.e., one that is producing X-rays with an energy of 75 keV. For such a machine, the emitted X-rays will have a frequency of 1.81×10^{19} Hz and a wavelength of 1.66×10^{-11} meters, or 0.166 Å (from Planck's Law). Clearly for such a source, it would be virtually impossible to make any measurements in the Near Field — i.e., within a very few wavelengths distant from the source — since even a six wavelength distance would be only 1 Å away (a 1 Å distance is less than the diameter of a methane molecule).

Measurements made in the Far Field of the strength or intensity of a radiating source then will always be **radiation** measurements, usually in units such as millirem/hour — mRem/hr. As stated earlier, **radiation** behaves according to the inverse square law, a relationship that states that radiation intensity decreases as the square of the distance between the point of measurement and the source.

Sources of Ionizing Radiation

Radioactivity

Radioactivity is the process by which certain unstable atomic nuclei undergo a nuclear disintegration. In this disintegration, the unstable nucleus will typically emit one or more of: (1) the common sub-atomic particles (i.e., the α-Particle, the β-Particles, etc.), and/or (2) photons of electromagnetic energy, (i.e., γ-Rays, etc.).

Radioactive Decay

Radioactive Decay refers to the actual process — involving one or more separate and distinct steps — by which some specific radioactive element, or radionuclide, undergoes the transition from its initial condition, as an "unstable" nucleus, ultimately to a later generation "unstable" radioactive nucleus, or — eventually — a "stable" nonradioactive nucleus.

In the process of this **Radioactive Decay**, the originally unstable nucleus will very frequently experience a change in its basic atomic number. Whenever this happens, its chemical identity will change — i.e., it will become an isotope of a different element. As an example, if an unstable nucleus were to emit an electron (i.e., a β^--particle), its atomic number would increase by one — i.e., an unstable isotope of calcium decays by emitting an electron, and in so doing becomes an isotope of scandium, thus:

$$_{20}^{45}\text{Ca} \rightarrow {}_{21}^{45}\text{Sc} + {}_{-1}^{0}\text{e},$$

which could also be written as follows:

$$_{20}^{45}\text{Ca} \rightarrow {}_{21}^{45}\text{Sc} + \beta^-$$

A second example would be the **Radioactive Decay** of the only naturally occurring isotope of thorium, which involves the emission of an α-particle:

$$_{90}^{232}\text{Th} \rightarrow {}_{88}^{228}\text{Ra} + {}_{2}^{4}\text{He},$$

which could also be written as follows:

$$_{90}^{232}\text{Th} \rightarrow {}_{88}^{228}\text{Ra} + {}_{2}^{4}\alpha$$

In this situation, the unstable thorium isotope was converted into an isotope of radium.

Radioactive Decay can occur in any of nine different modes. These nine are listed below, in each case with an example of a radioactive isotope that undergoes radioactive decomposition — in whole or in part — following the indicated decay mode:

Decay Mode	*Example*
Alpha Decay (α-decay)	$^{235}_{92}U \rightarrow {}^{231}_{90}Th + {}^{4}_{2}He$
Beta Decay (β^--decay)	$^{90}_{38}Sr \rightarrow {}^{90}_{39}Y + {}^{0}_{-1}e$
Positron Decay (β^+-decay)	$^{22}_{11}Na \rightarrow {}^{22}_{10}Ne + {}^{0}_{+1}e + \gamma$
	(simultaneous β^+ **and** γ-decay)
Gamma Decay (γ-decay)	$^{60}_{27}Co \rightarrow {}^{60}_{28}Ni + {}^{0}_{-1}e + \gamma$
	(simultaneous β **and** γ-decay)
Neutron Decay (n-decay)	$^{252}_{98}Cf \rightarrow {}^{107}_{42}Mo + {}^{141}_{56}Ba + 4{}^{1}_{0}n$
	(simultaneous n-decay **and** SF)
Electron Capture (EC)	$^{125}_{53}I + {}^{0}_{-1}e \rightarrow {}^{125}_{52}Te + \gamma$
	(simultaneous EC **and** γ-decay)
Internal Conversion (IC)	$^{125}_{52}Te \rightarrow {}^{0}_{-1}e$
	(following the simultaneous EC **and** γ-decay reaction shown above; the electron is ejected — i.e., IC — from one of the technetium atom's innermost electron sub-shells leaving the technetium atom positively ionized)
Isomeric Transition (IT)	$^{121m}_{50}Sn \rightarrow {}^{121}_{50}Sn + \gamma$
	(simultaneous IT **and** γ-decay)
Spontaneous Fission (SF)	$^{252}_{98}Cf \rightarrow {}^{107}_{42}Mo + {}^{141}_{56}Ba + 4{}^{1}_{0}n$
	(simultaneous SF **and** n-decay)

Radioactive Decay Constant

The **Radioactive Decay Constant** is the isotope specific "time" coefficient that appears in the exponent term of Equation 7.4. Equation 7.4 is the widely used relationship that always serves as the basis for determining the quantity (atom count or mass) of any as yet undecayed radioactive isotope. This exponential relationship is used to evaluate remaining quantities at any time interval after a starting determination of an "initial" quantity. By definition, all radioactive isotopes decay over time, and the **Radioactive Decay Constant** is an empirically determined factor that effectively reflects the speed at which the decay process has occurred or is occurring.

Mean Life

The **Mean Life** of any radioactive isotope is simply the average "lifetime" of a single atom of that isotope. Quantitatively, it is the reciprocal of that nuclide's Radioactive Decay Constant — see Equation 7.6. **Mean Lives** can vary over extremely wide ranges of time; as an example of this wide variability, the following are the **Mean Lives** of two fairly common radioisotopes, namely, the most common naturally occurring isotope of uranium and a fairly common radioactive isotope of beryllium:

For an atom of $^{238}_{92}U$, the Mean Life (_-decay) is 6.44×10^9 years

For an atom of ^7_4Be, the **Mean Life** (EC decay) is 76.88 days

---

Half-Life

The **Half-Life** of any radioactive species is the time interval required for the population of that material to be reduced, by radioactive decay, to one half of its initial level. The **Half-Lives** of different isotopes, like their Mean Lives, can vary over very wide ranges. As an example, for the two radioactive decay schemes described under the definition of Radioactive Decay, the **Half-Lives** are as follows

For $^{45}_{20}Ca$, the Half-Life is 162.7 days

For $^{232}_{90}Th$, the **Half-Life** is 1.4×10^{10} years

As can be seen from these two **Half-Lives**, this parameter can assume values over a very wide range of times. Although the thorium isotope listed above certainly has a very long **Half-Life**, it is by no means the longest. On the short end of the scale, consider another thorium isotope, $^{218}_{90}Th$, which has a **Half-Life** of 0.11 microseconds.

Nuclear Fission

Nuclear Fission, as the process that will be described here, differs from the Spontaneous Fission mode that was listed under the description of Radioactive Decay as one of the nine radioactive decay modes. This class of **Nuclear Fission** is a nuclear reaction in which a fissile isotope — i.e., an isotope such as $^{235}_{92}U$ or $^{239}_{94}Pu$ — upon absorbing a free neutron undergoes a fracture which results in the conversion of the initial isotope into:

1. two daughter isotopes,
2. two or more additional neutrons,
3. several very energetic γ-rays, and
4. considerable additional energy, usually appearing in the form of heat.

Nuclear Fission reactions are the basic energy producing mechanisms used in every nuclear reactor, whether it is used to generate electric power, or to provide the motive force for a nuclear submarine. One of the most important characteristics of this type of reaction is that by regenerating one or more of the particles (i.e., neutrons) that initiated the process, the reaction can become self-sustaining. Considerable value can be derived from this process if the chain reactions involved can be controlled. In theory, control of these chain reactions occurs in such things as nuclear power stations. An example of an uncontrolled **Nuclear Fission** reaction would be the detonation of an atomic bomb.

An example of a hypothetically possible **Nuclear Fission** reaction might be:

$$^{235}_{92}U + ^{1}_{0}n \rightarrow ^{109}_{44}Ru + ^{123}_{48}Cd + 3\,^{1}_{0}n + 3\,\gamma + \text{considerable energy}$$

In this hypothetical fission reaction, the sum of the atomic masses of the two reactants to the left of the arrow is 236.052589 amu, whereas the sum of atomic masses of all the products to the right of this arrow is 234.856015 amu. Clearly there is a mass discrepancy of 1.196574 amu or 1.987×10^{-24} grams. It is this mass that was converted into the several γ-rays that were created and emitted, as well as the very considerable amount of energy that was liberated. It appears that Albert Einstein was correct: mass and energy are simply different forms of the same thing.

Since **Nuclear Fission** reactions are clearly sources for a considerable amount of ionizing radiation, they are of interest to occupational safety and health professionals.

Radiation Measurements
The Strength or Activity of a Radioactive Source

The most common measure of **Radiation Source Strength** or **Activity** is the number of radioactive disintegrations that occur in the mass of radioactive material per unit time. There are several basic units that are employed in this area; they are listed below, along with the number of disintegrations per minute that each represents:

Unit of Source Activity	Abbreviation	Disintegrations/min
1 Curie	Ci	2.22×10^{12}
1 Millicurie	mCi	2.22×10^{9}
1 Microcurie	μCi	2.22×10^{6}
1 Picocurie	pCi	2.22
1 Becquerel	Bq	60

Exposure

Exposure is a unit of measure of radiation that is currently falling into disuse. The basic definition of **Exposure** — usually designated as X — is that it is the sum number of all the ions, of either positive or negative charge — usually designated as ΣQ — that are produced in a mass of air — which has a total mass, Σm — by some form of ionizing radiation that, in the course of producing these ions, has been totally dissipated. Quantitatively, it is designated by the following formula:

$$X = \frac{\sum Q}{\sum m}$$

The unit of **Exposure** is the roentgen, or R. There is no *SI* unit for **Exposure**; thus as stated above this measure is now only rarely encountered. References to **Exposure** are now only likely to be found in older literature.

Dose

Dose, or more precisely **Absorbed Dose**, is the total energy imparted by some form of ionizing radiation to a known mass of matter that has been exposed to that radiation. Until the mid 1970s the most widely used unit of **Dose** was the rad, which has been defined to be equal to 100 ergs of energy absorbed into one gram of matter. Expressed as a mathematical relationship:

$$1.0 \text{ rad} = 100 \frac{\text{ergs}}{\text{gram}} = 100 \text{ ergs} \cdot \text{grams}^{-1}$$

At present, under the *SI System*, a new unit of **Dose** has come into use. This unit is the gray, which has been defined to be the deposition of 1.0 joule of energy into 1.0 kilogram of matter. Expressed as a mathematical relationship:

$$1.0 \text{ gray} = 1.0 \frac{\text{joule}}{\text{kilogram}} = 1.0 \text{ joule} \cdot \text{kilogram}^{-1}$$

The gray is steadily replacing the rad although the latter is still in fairly wide use. For reference, 1 gray = 100 rad (1 Gy = 100 rad), or 1 centigray = 1 rad (1 cGy = 1 rad). For most applications, Doses will be measured in one of the following "sub-units": (1) millirad — mrads; (2) microrads — μrads; (3) milligrays — mGys; or (4) micrograys — μGys. These units are — as their prefixes indicate — either 10^{-3} or 10^{-6} multiples of the respective basic Dose unit.

Dose, as a measurable quantity, is always represented by the letter "D".

Dose Equivalent

The **Dose Equivalent** is the most important measured parameter insofar as the overall subject of radiation protection is concerned. It is basically the product of the Absorbed Dose and an appropriate Quality Factor, a coefficient that is dependent upon the type of ionizing particle involved — see Equation 7.12. This parameter is usually represented by the letter "H". There are two cases to consider, and they are as follows:

1. If the Dose or Absorbed Dose, D, has been given in units of rads (or mrads, or μrads), then the units of the Dose Equivalent, H, will be rem (or mrem, or μrem) as applicable.

2. If the Dose or Absorbed Dose, D, has been given in units of grays (or mGy, or μGy), then the units of the Dose Equivalent, H, will be sieverts (or mSv, or μSv) as applicable.

It is very important to note that since **1 Gray = 100 rads**, it follows that **1 sievert = 100 rem**.

Finally, if it is determined that a Dose Equivalent > 100 mSv, there is almost certainly a very serious situation with a great potential for human harm; thus, in practice, for Dose Equivalents above this level, the unit of the sievert is rarely, if ever, employed.

Relevant Formulae and Relationships
Basic Relationships for Electromagnetic Radiation
Equation 7.1

For any photon that is a part of the overall electromagnetic spectrum, the relationship between that photon's wavelength, its frequency, and/or its wavenumber is given by the following expression, Equation 7.1, which is shown below in two equivalent forms:

$$c = \lambda \nu$$
$$c = \frac{\nu}{k}$$

Where:

c = the speed of light in a vacuum, which is 2.99792458×10^8 meters/second (frequently approximated as 3.0×10^8 meters/second);

λ = the wavelength of the photon in question, in units of meters (actually meters/cycle);

ν = the frequency associated with the photon in question, in units of reciprocal seconds — sec^{-1} — (actually cycles/second or Hertz); and

k = the wavenumber of the photon in question, in units of reciprocal meters — $meters^{-1}$ — (actually cycles/meter)

Equation 7.2

The relationship between the wavelength and the wavenumber of any electromagnetic photon is given by the following expression, Equation 7.2:

$$\lambda = \frac{1}{k}$$

Where: λ =   the wavelength of the photon in question, in units of meters (actually meters/cycle), as defined above for Equation 7.1; and

          k =   the wavenumber of the photon in question, in units of reciprocal meters — meters^{-1} — (actually cycles/meter), also as defined for Equation 7.1.

Note: Wavenumbers very frequently are expressed in units of reciprocal centimeters — cm^{-1} — and when expressed in these units, the photon is said to be at "xxx" wavenumbers (i.e., a 3,514 cm^{-1} photon is said to be at 3,514 wavenumbers).

Equation 7.3

Equation 7.3 expresses the relationship between the energy of any photon in the electromagnetic spectrum, and the wavelength of that photon. This relationship is Planck's Law, which was the first specific, successful, quantitative relationship ever to be applied in the area of quantum mechanics. This Law, as the first significant result of Planck's basic research in this area, formed one of the main foundation blocks upon which modern physics and/or quantum mechanics was built.

$$E = h\nu$$

Where: E =   the energy of the electromagnetic photon in question, in some suitable energy unit — i.e., joules, electron volts, etc.;

          h =   Planck's Constant, which has a value of 6.626×10^{-34} joule·seconds, and/or 4.136×10^{-15} electron volt·seconds; and

          ν =   the frequency associated with the photon in question, in units of reciprocal seconds (actually cycles/second or Hertz) — as defined for Equation 7.1.

Calculations Involving Radioactive Decay
Equation 7.4

For any radioactive isotope, the following Equation, 7.4, identifies the current **Quantity** or amount of the isotope that would be present at any incremental time period after the initial or starting mass or number of atoms had been determined (i.e., the mass or number of atoms that has not yet undergone radioactive decay). With any radioactive decay, the number of disintegrations or decays per unit time will be exponentially proportional to both the Radioactive Decay Constant for that nuclide, and the actual numeric count of the nuclei that are present (i.e., the **Quantity**).

$$N_t = N_0 e^{-kt}$$

Where: N_t = the **Quantity** of any radioactive isotope present at any time, **t**; this **Quantity** is usually measured either in mass units (mg, µg, etc.) OR as a specific numeric count of the as yet undecayed nuclei remaining in the sample (i.e., 3.55×10^{19} atoms);

N_0 = the **Initial Quantity** of that same radioactive isotope — i.e., the **Quantity** that was present at the time, $t = t_0$ (i.e., 0 seconds, 0 minutes, 0 hours, 0 days, or whatever unit of time is appropriate to the units in which the Radioactive Decay Constant has been expressed). This is the "Starting" or **Initial Quantity** of this isotope, and it is always expressed in the same units as N_t, which is described above;

k = the **Radioactive Decay Constant**, which measures number of nuclear decays per unit time; in reality, the "number of nuclear decays" is a simple integer, and as such, is effectively dimensionless; thus this parameter should be thought of as being measured in reciprocal units of time (i.e., seconds^{-1}, minutes^{-1}, hours^{-1}, days^{-1}, or even years^{-1}, etc.); and

t	=	the **Time Interval** that has passed since the Initial Quantity of material was determined. This **Time Interval** must be expressed in an appropriate unit of time — i.e., the units of "**k**" and "**t**" must be mutually consistent; thus the units of "**k**" must be: seconds, minutes, hours, days, years, etc.

Equation 7.5

The following Equation, 7.5, provides the relationship between the **Half-Life** of a radioactive isotope and its **Radioactive Decay Constant**. The **Half-Life** of any radioactive nuclide is the statistically determined time interval required for exactly half of the isotope to decay, effectively leaving the other half of the isotope in its original form.

$$T_{1/2} = \frac{0.693}{k},$$

or

$$k = \frac{0.693}{T_{1/2}}$$

Where: $T_{1/2}$ = the **Half-Life** of the radioactive isotope under consideration; this parameter must be expressed in the same units of time that are used as reciprocal time units for the Radioactive Decay Constant; and

k = the **Radioactive Decay Constant**, measured in reciprocal units of time (i.e., seconds^{-1}, minutes^{-1}, hours^{-1}, days^{-1}, or even years^{-1}, etc.), as defined for Equation **7.4**.

Equation 7.6

The **Mean Life** of any radioactive isotope is the measure of the average 'lifetime'' of a single atom of that isotope. It is simply the reciprocal of that nuclide's Radioactive Decay Constant. Equation 7.6 provides the quantitative relationship that is involved in calculating this parameter.

$$\tau = \frac{1}{k} = \frac{T_{1/2}}{0.693} = 1.443T_{1/2}$$

Where: τ = the **Mean Life** of some specific radionuclide, expressed in units of

		time (i.e., seconds, minutes, hours, days, or years, etc.)
k	=	the **Radioactive Decay Constant**, measured in consistent reciprocal units of time (i.e., seconds^{-1}, minutes^{-1}, hours^{-1}, days^{-1}, or even years^{-1}, etc.); and
$T_{1/2}$	=	the **Half-Life** of the radioactive isotope under consideration; this parameter must be expressed in the same units of time as the Mean Life, and as the reciprocal of the time units in which the Radioactive Decay Constant is expressed.

Equations 7.7 and 7.8

The **Activity** of any radioisotope is defined to be the number of radioactive disintegrations that occur per unit time. equations 7.7 and 7.8 are two simplified forms of the relationship that can be used to calculate the **Activity** of any radioactive nuclide.

Equation 7.7

$$A_b = kN$$

Equation 7.8

$$A_c = \frac{kN}{3.70 \times 10^{10}} = \left[2.703 \times 10^{-11}\right]kN$$

Where: A_b	=	the **Activity** of the radionuclide, expressed in becquerels,
		or
A_c	=	the **Activity** of the radionuclide, expressed in curies;
k	=	the **Radioactive Decay Constant**, measured in reciprocal units of time (i.e., seconds^{-1}, minutes^{-1}, hours^{-1}, days^{-1}, or even years^{-1}, etc.); and
N	=	the **Quantity** of the radioactive isotope that is present in the sample at the time when the evaluation of the **Activity** is to be made, measured as a specific numeric count of the as yet

undecayed nuclei remaining in the sample (i.e., 3.55×10^{19} atoms);

Equations 9 and 10

The following two Equations, 7.9 and 7.10, provide the two more general forms of the relationship for determining the **Activity** of any radioactive nuclide.

Equation 7.9

$$A_t = kN_0 e^{-kt}$$

Equation 7.10

$$A_t = \left[\frac{0.693}{T_{1/2}}\right] N_0 e^{-(0.693)t/T_{1/2}}$$

Where: A_t =   the **Activity** of any radioactive nuclide at any time, **t**. The units of this calculated parameter will be becquerels;

k =   the **Radioactive Decay Constant**, measured in reciprocal units of time (i.e., seconds^{-1}, minutes^{-1}, hours^{-1}, days^{-1}, or even years^{-1}, etc.);

N_0 =   the **Initial Quantity** of that same radioactive isotope — i.e., the **Quantity** that was present at the time, $t = t_0$ (i.e., 0 seconds, 0 minutes, 0 hours, 0 days, or zero of whatever unit of time is appropriate to the dimensionality in which the Radioactive Decay Constant has been expressed) — this is the "Starting" or **Initial Quantity** of this isotope, measured as a specific numeric count of the as yet undecayed nuclei remaining in the sample (i.e., 3.55×10^{19} atoms);

$T_{1/2}$ =   the **Half-Life** of the radioactive isotope under consideration; this parameter must be expressed in the same units of time that appear as reciprocal time units for the Radioactive Decay Constant; and

t = the **Time Interval** that has passed since the Initial Quantity of material was determined; this **Time Interval** must be expressed in an appropriate unit of time — i.e., the units of "**k**" and "**t**" must be consistent with each other.

Dose and/or Exposure Calculations
Equation 7.11

The following Equation, 7.11, is applicable only to **Dose Exposure Rates** caused by high energy X-rays and/or γ-Rays (as well as — hypothetically, at least, but certainly not practically — any other photons such as a Cosmic Ray, which have a still shorter wavelength). Determinations of these **Dose Exposure Rates** are largely limited to medical applications. In order to be able to make these determinations, some very specific and unique source-based radiological data (i.e., the Radiation Constant of the source) must be known. In addition, the Radiation Source Activity, and the distance from the source to the point at which **Dose Exposure Rate** is to be measured, must also be known.

$$E = \frac{\Gamma A}{d^2}$$

Where: E = the **Dose Exposure Rate** that has resulted from an individual's exposure to some specific X- or γ-radiation source, for which the specific Radiation Constant, Γ, is known; this dose rate is commonly expressed in units such as Rads/hour;

Γ = the **Radiation Constant** for the X-ray or γ-Ray active nuclide being considered, expressed in units of (Rads · centimeters)^2 per millicurie · hour, or

$$\left[\frac{Rad \cdot cm^2}{mCi \cdot hr} \right];$$

A = the **Radiation Source Activity**, measured usually in millicuries (mCi's); and

d   =   the **Distance** between the "Target"
             and the radiation source, measured in
             centimeters (cm).

Equation 7.12

This Equation, 7.12, provides for the conversion of an **Absorbed Radiation Dose**, expressed either in Rads or in Grays, to a more useful form — useful from the perspective of measuring the magnitude of the overall impact of the dose on the individual who has been exposed. This alternative, and more useful, form of Radiation Dose is called the **Dose Equivalent** and is expressed either in **rems** or in **sieverts**, both of which measure the "Relative Hazard" caused by the energy transfer that results from an individual's exposure to various different types or categories of radiation. The **rem** and/or the **sievert**, therefore, is dependent upon two specific factors: (1) the specific type of radiation that produced the exposure, and (2) the amount or physical dose of the radiation that was involved in the exposure.

To make these determinations, a "Quality Factor" is used to adjust the measurement that was made in units of **rads** or **grays** — both of which are independent of the radiation source — into an equivalent in **rems** and/or **sieverts**.

This Quality Factor (QF) is a simple multiplier that adjusts for the effective *Linear Energy Transfer* (*LET*) that is produced on a target by each type or category of radiation. The higher the *LET*, the greater will be the damage that can be caused by the type of radiation being considered; thus, this alternative **Dose Equivalent** measures the overall biological effect, or impact, of an otherwise "simple" measured Radiation Dose.

The "range" of β- and/or α-rays is, as stated earlier, very limited — i.e., the "range" is the distance that any form of radiation is capable of traveling through solid material, such as metal, wood, human tissue, etc. before it is stopped. Because of this, Quality Factors as they apply to alpha and beta particles are only considered from the perspective of internal **Dose Equivalent** problems. Quality factors for neutrons, X-, and γ-rays apply both to internal and external **Dose Equivalent** situations.

$$H_{Rem} = D_{Rad}\left[QF\right]$$
and
$$H_{Sieverts} = D_{Grays}\left[QF\right]$$

Where: $\mathbf{H_{Rem}}$ or $\mathbf{H_{sievert}}$ =   the adjusted **Dose Equivalent** in the more useful "effect related" form, measured in either rems or sieverts (SI Units);

$\mathbf{D_{Rad}}$ or $\mathbf{D_{Gray}}$ =   the **Absorbed Radiation Dose**, which is independent of the type of radiation, and is measured in rads or grays (SI Units); and

QF = the **Quality Factor**, which is a properly dimensioned coefficient — either in units of rems/rad or sieverts/gray, as applicable — that is, itself, a function of the type of radiation being considered (Table 2).

Table 2. Tabulation of Quality Factors (QFs) by Radiation Type

Types of Radiation	*Quality Factors — QFs*	*Internal/External*
X-rays *or* γ-rays	1.0	Both
β-Rays (positrons *or* electrons)	1.0	Internal Only
Thermal Neutrons	5.0	Both
Slow Neutrons	4.0–22.0	Both
Fast Neutrons	3.0–5.0	Both
Heavy, Charged Particles (Alphas, etc.)	20.0	Internal Only

Calculations Involving the Reduction of Radiation Intensity Levels

Equation 7.13

This Equation, 7.13, identifies the effect that shielding materials have in reducing the intensity level of a beam of ionizing radiation. The **Radiation Emission Rate** produced by such a beam can be reduced either by interposing shielding materials between the radiation source and the receptor, or by increasing the source-to-receptor distance. Obviously, the **Radiation Emission Rate** could be decreased still further by using both approaches simultaneously. The approach represented by Equation 7.13 deals solely with the use of shielding materials (i.e., it does not consider the effect of increasing source-to-receptor distances). This approach involves the use of the Half-Value Layer (HVL) concept. A **Half-Value Layer** represents the thickness of any shielding material that would reduce, by one half, the intensity level of incident X- or γ-radiation. This expression is provided in two forms:

$$ER_{goal} = \frac{ER_{source}}{2^{\sqrt{HVL}}}$$

or

$$x = \frac{\log\left[\dfrac{ER_{source}}{ER_{goal}}\right][HVL]}{\log 2} = 3.32\log\left[\frac{ER_{source}}{ER_{goal}}\right][HVL]$$

Where: ER_{goal} =   the target **Radiation Emission Rate**, measured in units of radiation dose per unit time (i.e., Rads/hour);

          ER_{source} =   the observed **Radiation Emission Rate** to be reduced by interposing Shielding Materials, in the same units as ER_{goal};

          x   =   the **Thickness** of shielding material required to reduce the measured **Radiation Emission Rate** to the level desired, usually measured in units of centimeters or inches (cm or in); and

          HVL   =   the **Half-Value Thickness** of the Shielding Material being evaluated (i.e., the **Thickness** of this material that will halve the Intensity Level of incident X- or γ-radiation), measured in the same units as "**x**", above.

Equation 7.14

The following Equation, 7.14, is the relationship that describes the effect of increasing the distance between a point source of X- or γ-radiation and a receptor, as an alternative method for decreasing the incident radiation intensity on the receptor. The relationship involved is basically geometric, and is most commonly identified or referred to as **The Inverse Squares Law**.

$$\frac{ER_a}{ER_b} = \frac{S_b^2}{S_a^2}$$

$$or$$

$$ER_a S_a^2 = ER_b S_b^2$$

Where: ER_a =   the **Radiation Emission Rate**, or **Radiation Intensity**, in units of radiation dose per unit time (i.e., Sieverts/hour), measured at a distance, "**a**" units from the radiation source;

          ER_b =   the **Radiation Emission Rate**, or **Radiation Intensity**, in the same units as, ER_a, above, measured at a different distance, "**b**" units from the radiation source;

S_a = the "**a**" **Distance**, or the distance between the radiation source and the first position of the Receptor; this distance is measured in some appropriate unit of length (i.e., meters, feet, etc.); and

S_b = the "**b**" **Distance**, or the distance between the radiation source and the second — usually more distant — position of the Receptor; this distance is also measured in some appropriate unit of length, and most importantly in the same units of length as S_a, above (i.e., meters, feet, etc.).

Calculations Involving Optical Densities
Equation 7.15

The following Equation, 7.15, describes the relationship between the absorption of monochromatic visible light (i.e., laser light), and the length of the path this beam of light must follow through some absorbing medium. This formula relies on the fact that each incremental thickness of this absorbing medium will absorb the same fraction of the incident radiation as will each other identical incremental thickness of this same medium.

The logarithm of the ratio of the **Incident Beam Intensity** to the **Transmitted Beam Intensity** is used to calculate the **Optical Density** of the medium. This relationship, then, is routinely used to determine the intensity diminishing capabilities (i.e., the **Optical Density**) of the protective goggles that must be worn by individuals who must operate equipment that makes use of high intensity monochromatic light sources, such as lasers.

$$OD = \log\left[\frac{I_{incident}}{I_{transmitted}}\right]$$

Where: **OD** = the measured **Optical Density** of the material being evaluated, this parameter is dimensionless;

$I_{incident}$ = the **Incident Laser Beam Intensity**, measured in units of power/unit area (i.e., W/cm^2); and

$I_{transmitted}$ = the Transmitted Laser Beam Intensity, measured in the same units as $I_{incident}$, above.

Relationships Involving Microwaves
Equation 7.16

The following Equation, 7.16, provides the necessary relationship for determining the **Distance to the *Far Field*** for any radiating circular microwave antenna. The *Far Field* is that region that is sufficiently distant (i.e., more than 2 or 3 wavelengths away) from the radiating antenna, that there is no longer any interaction between the electrical and the magnetic fields being produced by this source. In the *Near Field* the interactions between the two electromagnetic fields being produced by any source require a different approach to the measurement of the effects, etc. The *Near Field* is every portion of the radiation field that is not included in the *Far Field* — i.e., it is that area that is closer to the source antenna than is the *Far Field*.

$$r_{FF} = \frac{A}{2\lambda} = \frac{\pi D^2}{8\lambda}$$

Where: r_{FF} = the **Distance to the *Far Field*** from the microwave radiating antenna (all distances equal to or greater than r_{FF} are considered to be in the *Far Field*; all distances less than this value will be in the *Near Field*), these distances are usually measured in centimeters (cm);

A = the **Area** of the radiating circular antenna, measured in square centimeters (cm^2) — for reference, this area can be calculated according to the following relationship,

$$\text{Circular Area} = \frac{\pi D^2}{4};$$

D = the circular microwave antenna **Diameter**, measured in centimeters (cm); and

λ = the **Wavelength** of microwave energy being radiated by the circular antenna, also measured in centimeters (cm).

Equation 7.17

The following Equation, 7.17, provides the relationship for determining the *Near Field* **Microwave Power Density** levels that are produced by a circular microwave antenna, radiating at a known **Average Power Output**.

$$W_{NF} = \frac{4P}{A} = \frac{16P}{\pi D^2}$$

Where: W_{NF} = the *Near Field* **Microwave Power Density**, measured in milliwatts/cm^2 (mW/cm^2);

P = the **Average Power Output** of the microwave radiating antenna, measured in milliwatts (mW);

A = the **Area** of the radiating circular antenna, measured in square centimeters (cm^2) — for reference, this area can be calculated according to the following relationship,

$$\text{Circular Area} = \frac{\pi D^2}{4}; \text{ and}$$

D = the circular microwave antenna **Diameter**, measured in centimeters (cm).

Equations 7.18 and 7.19

The following two Equations, 7.18 and 7.19, provide the basic approximate relationships that are used for calculating either microwave **Power Density Levels** in the *Far Field* (Equation 7.18), OR, alternatively, for determining the actual *Far Field* **Distance** from a radiating circular microwave antenna at which one would expect to find some specific **Power Density Level** (Equation 7.19).

Unlike the Equation at the top of this page (i.e., Equation 7.17), these two formulae have been empirically derived; however, they may both be regarded as sources of reasonably accurate values for the **Power Density Levels** at points in the *Far Field* (Equation 7.18), or for various *Far Field* **Distances** (Equation 7.19).

Equation 7.18

$$W_{FF} = \frac{AP}{\lambda^2 r^2} = \frac{\pi D^2 P}{4\lambda^2 r^2}$$

Equation 7.19

$$r = \frac{1}{\lambda}\sqrt{\frac{AP}{W_{FF}}} = \frac{D}{2\lambda}\sqrt{\frac{\pi P}{W_{FF}}}$$

Where: W_{FF} =   the **Power Density Level** at a point in the *Far Field* that is "**r**" centimeters distant from the circular microwave antenna, with this Power Density Level measured in milliwatts/cm² (mW/cm^2);

r =   the *Far Field* Distance (from the point where the **Power Density Level** is being evaluated) to the radiating circular microwave antenna, also measured in centimeters (cm);

D =   the circular microwave antenna's **Diameter**, measured in centimeters (cm).

A =   the **Area** of the radiating circular antenna, measured in square centimeters (cm^2) — for reference, this area can be calculated according to the following relationship.

$$\text{Circular Area} = \frac{\pi D^2}{4};$$

λ =   the **Wavelength** of microwave energy being radiated by the circular antenna, also measured in centimeters (cm); and

P =   the **Average Power Output** of the microwave radiating antenna, measured in milliwatts (mW).

Chapter 8
Statistics and Probability

This chapter will discuss the broad areas of statistics and probability, as these disciplines can be applied to the routine practice of occupational safety and health. Decision making on matters of employee safety frequently involves the evaluation of statistical data, and the subsequent development from these data of the probabilities of the occurrence of future events. These evaluations and the subsequent projections are important because the events being considered may involve workplace hazards. These two subjects: (1) the statistical aspects and (2) the probability considerations will be considered separately.

Relevant Definitions
Populations

A **Population** is any set of values of some variable measure of interest — for example, a listing of *the orthodontia bills of every person living on the island of Guam,* or a tabulation showing the count of *the number of Letters to the Editor that were received by the Washington Post newspaper each day during 1996,* would each make up a **Population**. A **Population** is the entire set of those values, the entire family of objects, data, measurements, events, etc. being considered from a statistical, probabilistic, or combinatorial perspective. A **Population** may consist of "events" that are either random or deterministic. For reference, a deterministic event is one that can be characterized as "cause-and-effect" related — i.e., when a person loses his grip on a baseball (the "cause"), the ball will fall to the ground (the "effect" event that was deterministically produced in a totally predictable manner by the identified "cause"). **Populations** may also consist of "members" whose values are themselves functions of a second, or a third, or even some higher number of random variables. The two example **Populations** listed above are most likely random (and therefore, not deterministic) — i.e., in each case, the values in either of these **Populations** are not obviously related to, or functions of, any other identifiable random factor or variable.

Distributions

A **Distribution** is a special type or subset of a population. It is a population, the values of whose "members" are related or a function of some identifiable and quantifiable random variable. A **Distribution** is virtually always spoken of or characterized as being "a function of some random variable"; the most common mathematical way to represent such a **Distribution** is to speak of it as a function of "x" — i.e., $f(x)$, where "x" is the random variable. Examples of Distributions might be *the per acre yield of soybeans as a function of such things as: (1) the amount of fertilizer applied to* *the crop, (2) the volume of irrigation water used, (3) the average daytime temperature during the growing season, (4) the acidity of the soil,* etc. Any **Distribution** that is characterized as being an **f(x)**, for "x", some continuous random variable, can be and is also frequently described as being:

1. a Probability Density Function,
2. a Probability Distribution,
3. a Frequency Function, and/or
4. a Frequency Distribution, etc.

Specific Types of Distributions
Uniform Distribution

A **Uniform Distribution** is one in which the value of every member is the same as the value of every other member. An example of a **Uniform Distribution** would be *the situation where the Safety Manager of a manufacturing plant had to complete safety inspections of various production areas at random times during the 8-hour workday.* If this workday is thought of as being divided up into 480 one-minute intervals, the probability of the Safety Manager visiting during any one of these intervals will be equally likely. Clearly — if the Safety Manager actually makes his visits on a random basis — each of these intervals will be equally likely to be selected; thus the "value" for each of these intervals will be equal (i.e., the probability of a visit during any specific interval will be 1/480, or 0.00208), and the population of these values can be said to constitute a **Uniform Distribution**.

Normal Distribution

A **Normal Distribution** is one of the most familiar types in this overall category of distributions — its applications apply to virtually any naturally occurring event. The "graphical" representation of a **Normal Distribution** is the well-known and widely understood "bell-shaped curve", or "normal probability distribution curve". The **Normal Distribution** is almost certainly the most important and widely used foundation block in the science of statistical inference, which is the process of evaluating data for the purpose of making predictions of future events. This type of distribution is always perfectly symmetrical about its Mean, described later in this chapter. Examples of **Normal Distributions** are: (1) *the number of tomatoes harvested during one growing season from each plant in a one-acre field of this crop*; (2) *the annual rainfall at some specific location on the island of Kauai, HI*; (3) *the magnitude of the errors that arise in the process of reading a dial oven thermometer,* etc.

Binomial Distribution

A **Binomial Distribution** is one in which every included event will have only two possible outcomes. It is a distribution made up of members whose values depend upon a binomial random variable. This category of variable most easily can be understood by considering one of its most familiar members, namely, the result of flipping a coin — a process for which there are only two possible outcomes, "heads" and "tails" (we assume the coin cannot land on and remain on its edge). An example of a Binomial Distribution would be the genders of all the individuals standing in the Ticket Line for the musical, *Phantom of the Opera.* **Binomial Distributions** in general, and particularly those with a large number of members, can be considered and handled, for any necessary computational effort, as Normal Distributions.

Exponential Distribution

An **Exponential Distribution** is frequently described as the Waiting Time Distribution, since many populations in this category involve considerations of variable time intervals. This class of distribution is relatively easy to understand by considering a couple of examples. A first might be *the lengths of time between Magnitude 7.5+ earthquakes on the San Andreas Fault in California.* Another example might be *the distances traveled by a municipal bus between major mechanical breakdowns, etc.* Both of these populations would be characterized as **Exponential Distributions**.

Charactristics of Populations and/or Distributions
Member

A **Member** of any population or distribution is simply one item from the set that makes up the whole. The **Member** can be any quantifiable characteristic — i.e., the height of any individual who belongs to some social group; the number of shrimp caught each day by any member of the Freeport, TX, fishing fleet; the number of times that the dice total 12 in a game of Craps, etc.

Variable

A **Variable** is a characteristic or property of any individual member of a population or distribution. The name, "**Variable**", derives from the fact that any particular characteristic of interest may assume different values among the individual members of the population or distribution being considered. If one was considering the distribution of the weights of elephant calves born in captivity throughout the world, one might evaluate such data from a variety of different random perspectives, or from the relationship of these birth weights to a variety of **Variables**. Among such **Variables** might be: (1) the country in which the birth occurred, (2) whether or not the birth occurred in a zoo, (3) a situation where the calf was the offspring of a "work elephant," or (4) the age of the mother elephant, etc.

Sample

A **Sample** is a subset of the members of an entire population. **Samples**, per se, are employed whenever one must evaluate some measurable characteristic of the members of an entire population in a situation where it is simply not feasible to consider or measure every member of that population. For example, one might have to answer a question of the following type:

1. Does the average digital clock produced in a clock factory actually keep correct time? or

2. Is the butterfat content of the daily output of homogenized milk from a dairy at or above an established standard for this factor?

In order to make any of these types of determinations, it is not usually considered necessary to sample and test every member of the population — rather such a determination can usually be made by obtaining and testing a **Sample** from the population of interest. For the two questions asked above, one might sample and test one of every 10 clocks, or one of every 1,000 gallons of milk, etc.

Parameter

A **Parameter** is a *calculated quantitative measure* that provides a useful description or characterization of a population or distribution of interest. **Parameters** are calculated directly from observations, the summary tabulation of which make up the population or distribution being considered. For any population or distribution of interest, an example of a **Parameter** would be that population's or distribution's Mean or Median (i.e., see Page 4 for complete descriptions of these terms).

Sample Statistic

A **Sample Statistic** is a specific numeric descriptive measure of a sample. It is calculated directly from observations made on the sample itself. Basically, a **Sample Statistic** is a *parameter* that is determined for a sample — i.e., the sample standard deviation (see section later in this chapter for a complete description of this term). It is very common for a measured **Sample Statistic** to be thought of as representative of or applicable to the entire population or distribution of interest.

Parameters of Populations and/or Distributions
Frequency Distribution

A **Frequency Distribution** is a tabulation of any of variable characteristics of any population that can be measured, counted, tabulated, or correlated. For example, from the **Frequency Distribution** that represents the results of the performance of high school seniors on the Scholastic Aptitude Test, it can be predicted that a score of 1,290 will place the student in the top 5% of all similar students taking this test.

Range

The **Range** of any set of variable data — taken from some population or distribution of interest — will be the calculated result that is obtained when the value of the numerically smallest member of the set is subtracted from the value of the numerically largest member of that same set — see Equation 8.1.

Mean

The **Mean** of any set of variable data — from some population or distribution of interest — is the sum of the individual values of the items of that data set, divided by the total number of items that make up the set. The **Mean** is the average value for the set of data being considered, and, in fact, the word "Average" is almost always used synonymously with **Mean**. The **Mean** is the first important measure of the "central tendency" of that set of variables — see Equation 8.3.

Geometric Mean

The **Geometric Mean** is a common alternative measure of the "central tendency" of any set of variable data — from some population or distribution of interest. It is a somewhat more useful measure than the simple Mean for any situation where the population or distribution being evaluated has a very large range of values among its members — i.e., a range of values varying over several orders of magnitude. Specifically, for any set of data, for which the ratio $R \geq 200$ or $\log R \geq 2.30$ — where R is defined as follows:

$$R = \frac{\text{the numeric value of the largest member of a population or distribution of interest}}{\text{the numeric value of the smallest member of a population or distribution of interest}}.$$

The **Geometric Mean** may be a better measure of this population's or distribution's central tendency — See Equation 8.4.

Median

The **Median** of any set of variable data — taken from some population or distribution of interest — is the middlemost value of that data set. When all the individual variable members of the set have been arranged either in ascending or descending order, the **Median** will be either:

1. the data point that is exactly in the center position, or
2. if there are a number of same value data points at, near, or around the center position, then this parameter will be the value of the data point that is centermost.

It can be regarded as the "Midpoint" value in any Normal Distribution containing "n" different numeric values, x_i. For such a set, it is that specific value of $x_{n/2}$, for which there are as many values in the distribution greater than this number, as there are values in the distribution less than this number. It is the second important measure of the "central tendency" of the set of variables being considered — see Equation 8.5.

Mode

The **Mode** of any set of variable data points — taken from some population or distribution of interest — is the value of the most frequently occurring member of that set. The **Mode** is the "most populous" value in any Normal Distribution containing "n" different numeric values, x_i. For such a set, it is that specific x_i which is the most frequently occurring value in the entire distribution. The **Mode** is the third most important measure of the "central tendency" of the set of variables being considered; however, it does not have to be a value that is close to the center of that population. It can be numerically the smallest, or the largest, or any other value in the set, so long as it appears more frequently than any other value — see Equation 8.6.

Sample Variance

The **Sample Variance** of any set of "n" data points — taken from some population or distribution of interest — is equal to the sum of the squared distances of each member of that set from the set's Mean. This squared "distance" must then be divided by one less than "n", the number of members of that set — i.e., the denominator in this process is the quantity, "(n − 1)" — see Equation 8.7.

This parameter looks at the absolute "distance" between each value in the set and the value of the set's Mean. If one were simply to obtain a simple "average" of these distances, the result would be zero, since some of these values would be negative, while a compensating number would be positive. To correct for this in the computation of the **Sample Variance**, each of these "distances" is squared; thus the result for each of these operations will always be positive, and a measure of the absolute "value-to-mean distance" will thereby be obtained.

The **Sample Variance** is always designated by the term, "s^2", and its dimensions will always be the square of the dimensions of the values of the members of the population or distribution being considered — i.e., if the population is a set of values measured in U.S. Dollars, then s^2 will be in units of (U.S. dollars)^2. For a Normal Distribution, the **Sample Variance** will probably be the best and least biased (i.e., the most unbiased) estimator of the true Population Variance.

Sample Standard Deviation

The **Sample Standard Deviation** of any set of variable data points — taken from some population or distribution of interest — is equal to the positive square root of the Sample Variance, as defined above on this page. For the relationship that defines this parameter, see Equation 8.9.

The **Sample Standard Deviation** is always designated by the term, "s", and its dimensions will always be the same as the dimensions of each member in the population or distribution being considered — i.e., if the population is a set of values measured in U.S. dollars, then "s" (unlike the

Sample Variance, "s^2", of which "s" is the square root) will also be in units of U.S. dollars.

For a Normal Distribution, the **Sample Standard Deviation** will be a better, less biased estimator of the true and most useful Population Standard Deviation.

Sample Coefficient of Variation

The **Sample Coefficient of Variation** is simply the ratio of the Sample Standard Deviation to the Mean of or for the population or distribution being considered — see Equation 8.11. This parameter is also commonly described as the **Relative Standard Deviation**.

For any Normal Distribution, the **Sample Coefficient of Variation** is thought to be a good to very good measure of the specific dispersion of the values that make up the set being examined. This coefficient is most commonly designated as "CV_{sample}," and it is a dimensionless number. Since the **Sample Coefficient of Variation** is regarded as a less biased, and therefore better estimator of the dispersion that characterizes the data in the distribution being considered, and does so more effectively than does its more biased counterpart, the Population Coefficient of Variation, this parameter tends to be the much more widely used of the two.

Population Variance

The **Population Variance** of any set of "n" data points — taken from some population or distribution of interest — is equal to the average of the squared distances of each member of that set from the Mean of the set — see Equation 8.8.

This parameter, like its Sample Variance counterpart, also looks at the absolute "distance" between each value in the set and the value of the set's Mean. Again, if one were simply to obtain a simple "average" of these distances, the summation result would always be zero, since roughly half of these distances are negative, while the remainder are positive. To correct for this in this computation and thereby obtain a true measure of the absolute distance, each of these "distances" is squared; thus the result will always be a positive number, and a very effective measure of the absolute "value-to-mean distance" will thereby be obtained.

The **Population Variance** is always designated by the term, "σ^2", and its dimensions will always be the square of the dimensions of each member in the population being considered — i.e., if the population is a set of values measured in units of "lost time injuries/1,000 work days", then σ^2 will be in units of (lost time injuries/1,000 work days)^2.

For a Normal Distribution, the **Population Variance** will usually be slightly more biased in determining a useful and precise value for this parameter than will its Sample Variance counterpart, and for this reason, it is used less frequently than the Sample Variance.

Population Standard Deviation

The **Population Standard Deviation** of any set of variable data points — taken from some population or distribution of interest — is equal to the positive square root of the Population Variance, as defined above — see Equation 8.10 for the mathematical relationship for the **Population Standard Deviation**.

The **Population Standard Deviation** is always designated by the term, "σ", and its dimensions will always be the same as the dimensions of each value in the population being considered — i.e., if the population is a set of values measured in "lost time injuries/1,000 work days", then "σ" (unlike the Population Variance, of which "σ" is the square root) will also be in units of "lost time injuries/1,000 work days".

For a Normal Distribution, the **Population Standard Deviation** will be slightly more biased as an estimator; thus, it is used less frequently in these determinations than the Sample Standard Deviation.

Population Coefficient of Variation

The **Population Coefficient of Variation** is simply the ratio of the Population Standard Deviation to the Mean of or for the population or distribution being considered — see Equation 8.12.

For any Normal Distribution, the **Population Coefficient of Variation** is thought to be a slightly biased measure of the specific dispersion of the values that make up the set being examined. This coefficient is most commonly designated as "$CV_{population}$", and it is a dimensionless number. Since the **Population Coefficient of Variation** is regarded as a slightly more biased, and therefore poorer estimator of the dispersion that characterizes the data in the distribution being considered, its counterpart, the Sample Coefficient of Variation, tends to be much more widely used.

Probability Factors and Terms

Experiment

An **Experiment** is a procedure or activity that will ultimately lead to some identifiable outcome that cannot be predicted with certainty. A good example of an **Experiment** might be the result of throwing a fair die and observing the number of dots that appear on the up-face. There are six possible result outcomes for such an **Experiment**; in order they are: one dot, two dots, three dots, four dots, five dots, and six dots. Each of these outcomes is equally likely; however, the specific result of any single **Experiment** can never be predicted with certainty.

Result

A **Result** is the most basic and simple outcome of any Experiment — i.e., for the Experiment of throwing of a fair die, there are a total of six possible **Results**, as described above.

Sample Space

The **Sample Space** of any Experiment is the totality of all the possible Results of that Experiment. For the Experiment of throwing a fair die described above, the **Sample Space** would be: one, two, three, four, five, and six. This **Sample Space** is most frequently represented symbolically in the following way:

$$S: \{1, 2, 3, 4, 5, 6\}$$

Event

An **Event** is a sub-set of specific Results from some well-defined overall Sample Space — i.e., for the fair die throwing Experiment described above, a specific **Event** might be the occurrence of an even number on the up-face of the die. From the totality of the Sample Space for this Experiment, the even number on the up-face of the die **Event** would be the following sub-set: two, four, and six — or listing this **Event** as a sort of Sub-Sample Space, the following would be its symbolic representation:

$$S_{even}: \{2, 4, 6\}$$

Compound Event

A **Compound Event** is some useful or meaningful combination of two or more different Events. Compound Events are structured in two very specific ways. In order, these structures are shown below:

1. The *union* of two Events — say, M and N — is the first type of a **Compound Event**. A *union* is said to have taken place whenever either M or N, or both M and N occur as the outcome of a single execution of the Experiment. Symbolically, a UNION, as the first category of a **Compound Event**, is represented in the following way — again assume we are dealing with the two Events, M and N:

$$M \cup N$$

Considering again the Experiment of throwing a fair die and observing its up-face, we might have an interest in the following two events: (1) M = the Result is an even number, and (2) N = the Result is a number greater than three. The Sub-Sample Space that makes up the UNION of these two Events would be:

$$S_{M \cup N}: \{2, 4, 5, 6\}$$

2. The *intersection* of two Events — again, say, M and N — is the second type of **Compound Event**. An *intersection* is said to have taken place whenever both M and N occur as the outcome of a single execution of the Experiment. Symbolically, an *intersection*, as the second category

of a **Compound Event**, is represented in the following way — again assume we are dealing with the two Events, M and N:

$$M \cap N$$

Considering again the die throwing Experiment, and the same two events described above in the section on the UNION, the Sub-Sample Space that makes up the INTERSECTION of these two events would be:

$$S_{M \cap N} : \{4, 6\}$$

Complementary Event

A **Complementary Event** is the totality of all the alternatives to some specific Event of interest. Within any Sample Space, the **Complement** to some Event of interest — say, M — will be every other possible Result that is not included within M. That is to say, whenever M has not occurred, its **Complement** — designated symbolically as M′ — will have occurred.

Considering again the Experiment of throwing a fair die and observing its resultant up-face, we might have an interest in the event: M = the Result is an even number. For this event, its **Complement**, M′ = the Result, is an odd number. The Sub-Sample Spaces for the Event, M, would be shown symbolically as:

$$S_M: \{2, 4, 6\}$$

The Sub-Sample Space for the **Complement** to M, again designated as M′, would be:

$$S_{M'}: \{1, 3, 5\}$$

Probabilities Associated with Results

The **Probability of the Occurrence of a Result** must always lie between 0 and 100% (or as a decimal, between 0.00 and 1.00). This probability is a measure of the relative frequency of occurrence of the Result of interest. It is the outcome frequency that would be expected to occur if the Experiment were repeated over and over and over — i.e., a very large number of repetitions. For example, in the Experiment of throwing and observing the up-face of a fair die, the probability of observing a "two" would be 1/6. This 1/6 factor would also be the probability associated with each one of the other five Results that exist within this Experiment's Sample Space.

It is important to note in this context that the probabilities of all the Results within any Sample Space must always equal 100%, or 1.00.

Probability of the Occurrence of Any Type of Event

The **Probability of the Occurrence of any Type of Event** can be determined by following the following five-step process:

1. Define as completely as possible the **Experiment** — i.e., describe the process involved, the methodology of making observations, the way these observations will be documented, etc.
2. Identify and list all the possible individual experimental **Results**.
3. Assign a probability of occurrence to each of these **Results**.
4. Identify and document the specific **Results** that will make up or are contained in the **Event**, the **Compound Event**, or the **Complementary Event** of interest.
5. Sum up the Result probabilities to obtain the Probability of the Occurrence of the Event, the Compound Event, or the Complementary Event of interest.

Relevant Formulae and Relationships
Parameters Relating to Any Population or Distribution
Equation 8.1

The following Equation, 8.1, defines the **Range** for any data set, population, or distribution of interest. It is determined by subtracting the **Value of the Numerically Smallest Member** of the set from the **Value of the Numerically Largest Member**.

$$R = \left[x_{i_{maximum}} - x_{i_{minimum}} \right]$$

Where: **R** =   the **Range** of the data set, population, or distribution consisting of "**n**" different members designated as "x_i";

x_i =   any of the "**n**" members of the data set, population, or distribution being considered;

$i_{maximum}$ =   the subscript index of the numerically largest member of the data set, population, or distribution being considered — indicating in Equation 8.1 the numerically largest member of the set by the term:

$x_{i_{maximum}}$; and

$i_{minimum}$ =   the subscript index of the numerically largest member of the data set, population, or distribution being considered — indicating in Equation 8.1 the numerically smallest member of the set by the term:

$x_{i_{minimum}}$.

Equation 8.2

The relationship that is used to characterize the relative magnitude of the range for any data set, distribution, or population under consideration is given by Equation 8.2. This expression is simply the ratio of the numerically largest member of any data set to its smallest member. This ratio is used to characterize the magnitude of the range for any distribution, population, or data set. Whenever a distribution, population, or data produces a value for **R** that is greater than 200, that distribution, population, or data set is said to have a relatively large range.

$$ R = \frac{X_{i_{maximum}}}{X_{i_{minimum}}} $$

Where: **R**   =   the ratio of the largest member of any distribution or population to the smallest member of the same distribution or population;

$X_{i_{maximum}}$ =   is the Value of the largest member of the distribution or population under consideration; and

$X_{i_{minimum}}$ =   is the Value of the smallest member of the distribution or population under consideration.

Equation 8.3

The following Equation, 8.3, defines the first, and the most important and, almost certainly the most widely used measure of location — or "central tendency" — for any type of population, distribution, or data set. This measure has been identified under a variety of names, among which are: **Mean**, Average, Arithmetic Mean, Arithmetic Average, etc. For the purpose of discussion in this text from this point forward, this parameter will always be identified as the **Mean**. In general, the **Mean** is designated either by the Greek letter, "μ", or by "$\bar{x}$".

$$ \mu = \bar{x} = \frac{1}{n} \sum_{i=1}^{n} x_i $$

Where: $\mu = \bar{x}$ =   the **Mean** of the population, distribution, or data set of "**n**" different values of x_i — the dimensions of the **Mean** and the individual members in the population, distribution, or data set will always be identical;

$$x_i \quad = \quad \text{the value of the "ith" member of the total of "n" members in the overall population, distribution, or data set;}$$

$$n \quad = \quad \text{the number of members in the overall population, distribution, or data set being considered; and}$$

$$i \quad = \quad \text{the "index" of the population, distribution, or data set being considered, this term will always appear as a subscript on the term representing a variable member of the overall population, distribution, or data set; this index will identify the position of the subscripted member within the overall population, distribution, or data set.}$$

Equation 8.4

The following Equation, 8.4, characterizes and defines a second measure of location — or "central tendency" — for any measurable or quantifiable parameter, for any distribution (normal or otherwise). This measure is called the **Geometric Mean** of the distribution. It is somewhat more useful than the simple Mean — at least as a measure of this "central tendency" — whenever the distribution being examined or analyzed has a very large range, which might be defined as one with values varying over several orders of magnitude (i.e., a range for which $R \geq 200$, or $log R \geq 2.30$ — see Equation 8.2).

Whenever a distribution has such a large range, the **Geometric Mean** will probably be a better indicator of its "central tendency" than will the simple Mean. It must be noted, however, that one can determine a **Geometric Mean** value for any distribution, population, or data set regardless of the magnitude of its range.

The relationships that are used to calculate this parameter are given below in two forms: the first is simply the direct mathematical relationship representing the definition of the **Geometric Mean**, while the second is presented in a format that will probably prove to be slightly easier to use in any case where the value of this parameter must be determined — particularly, for any distribution that has a relatively large to very large range.

$$M_{geometric} = \sqrt[n]{(x_1)(x_2)(x_3)\ldots(x_{n-1})(x_n)}$$

$$M_{geometric} = 10^{\left[\frac{1}{n}\sum_{i=1}^{n} \log x_i\right]}$$

Where: $M_{geometric}$ =   the **Geometric Mean** of the
                                  distribution, population, or data set un-
                                  der consideration;

           x_i =   is the value of the "**ith**" of "**n**" members
                                  of the overall distribution, population, or
                                  data set under consideration;

           n   =   the number of members in the
                                  distribution, population, or data set
                                  under consideration.

Equation 8.5

The following Equation, 8.5, is actually more of a definition. It characterizes the third measure of location, or "central tendency", for any quantifiable parameter, preferably for the situation in which the information being analyzed makes up a normal distribution. This parameter is called the **Median**. Although it is considered to be most applicable to normal distributions, a **Median** value can be determined for any other type of distribution, population, or data set.

M_e = the **Median** or "midpoint" value (principally for a normal distribution) of "n" different numeric values of "x_i" — i.e., when all the members of the distribution, population, or data set have been arranged in an increasing or a decreasing order by their numeric values. the **Median** will be in the middle position of the resultant ordered set. If "**n**" is odd, then the **Median** will be the *actual middle number* in the data set. If "**n**" is even, then the **Median** will be the *numeric average*, or *mean*, of the two members of the ordered data set that jointly occupy the middle position of that set.

Where: M_e =   the **Median** of the distribution,
                                  population, or data set consisting of "**n**"
                                  different values of x_i;

           x_i =   is the value of the "ith" of "**n**" members
                                  of the overall distribution, population, or
                                  data set under consideration;

           n   =   the number of members in the overall
                                  distribution, population, or data set
                                  under consideration.

Equation 8.6

The following Equation, 8.6, is also more of a definition. It characterizes the fourth measure of location, or "central tendency", for any quantifiable parameter, again preferably for a situation in which the resultant distribution is normal. This parameter is called the **Mode**. Although it is considered to apply most effectively to normal distributions, the **Mode** can also be determined for any other type of distribution, population, or data set.

$M_o =$ the **Mode** or "most populous" value in any distribution, population, or data set consisting of "**n**" different numeric values of "x_i", i.e., that specific numeric value of "x_i" which is the most frequently occurring value in the entire distribution, population, or data set. Although the **Mode** is considered to be an important measure of location or "central tendency", this value can occur at any position in the data set — i.e., it could be the smallest value, or the largest, or any other value. In a normal distribution, the **Mode** will usually be fairly close in value to the Median, and therefore, this parameter will provide its most useful information when applied to this important class of distribution.

Where: M_o = the **Mode** of the distribution, population, or data set of "**n**" different Values of "x_i";

x_i = is the value of the "**ith**" of "**n**" members of the overall distribution, population, or data set under consideration;

n = the number of members in the overall distribution, population, or data set under consideration.

Equation 8.7

The following Equation, 8.7 is shown in two equivalent forms, and defines the **Sample Variance**, which is the first and most widely used measure of variability, or dispersion, of the data in any distribution, population, or data set of interest.

$$s^2 = \frac{\sum_{i=1}^{n}\left[x_i - \mu\right]^2}{n - 1} = \frac{\sum_{i=1}^{n}\left[x_i - \bar{x}\right]^2}{n - 1}$$

Where: s2 = the Sample Variance for the entire distribution, population, or data set of "n" different values of "xi";

xi = is the value of the "ith" of "n" members of the overall distribution, population, or data set under consideration;

n	=	the number of members in the overall distribution, population, or data set under consideration; and
$\mu = \bar{x}$	=	the Mean of the distribution, population, or data set.

Equation 8.8

The following Equation, 8.8, is shown in two equivalent forms, and defines the **Population Variance**, which is the second measure of variability, or dispersion, of the data in any distribution, population, or data set of interest.

$$\sigma^2 = \frac{\sum\limits_{i=1}^{n}\left[x_i - \mu\right]^2}{n} = \frac{\sum\limits_{i=1}^{n}\left[x_i - \bar{x}\right]^2}{n}$$

Where:	σ^2	=	the **Population Variance** for the entire distribution, population, or data set of "**n**" different values of "x_i";
	x_i	=	is the value of the "**ith**" of "**n**" members of the overall distribution, population, or data set under consideration;
	n	=	the number of members in the overall distribution, population, or data set under consideration; and
	$\mu = \bar{x}$	=	the Mean of the distribution, population, or data set.

Equation 8.9

The following Equation, 8.9, which like its two predecessors is shown in two equivalent forms, defines the **Sample Standard Deviation**, which is the third — and probably most important — measure of variability, or dispersion, of the data in any distribution, population, or data set of interest. In general, the **Sample Standard Deviation** is believed to be most applicable to normal distributions; however it can be and is applied to any type of data set.

$$s = \sqrt{s^2} = \sqrt{\frac{\sum\limits_{i=1}^{n}\left[x_i - \mu\right]^2}{n-1}} = \sqrt{\frac{\sum\limits_{i=1}^{n}\left[x_i - \bar{x}\right]^2}{n-1}}$$

Where: s   =   the **Sample Standard Deviation** for the entire distribution, population, or data set of "**n**" different values of "x_i";

s^2 =   the **Sample Variance** for the entire distribution, population, or data set of "**n**" different values of "x_i";

x_i =   is the value of the "**ith**" of "**n**" members of the overall distribution, population, or data set under consideration;

n   =   the number of members in the overall distribution, population, or data set under consideration; and

$\mu = \bar{x}$ =   the Mean of the distribution, population, or data set.

Equation 8.10

The following Equation, 8.10, which like its three predecessors is shown in two equivalent forms, defines the **Population Standard Deviation**, which is the fourth measure of variability, or dispersion, of the data in any distribution, population, or data set of interest. In general, the **Population Standard Deviation** is believed to be the least important of the variability or dispersion quantifying parameters.

$$\sigma = \sqrt{\sigma^2} = \sqrt{\frac{\sum_{i=1}^{n}\left[x_i - \mu\right]^2}{n}} = \sqrt{\frac{\sum_{i=1}^{n}\left[x_i - \bar{x}\right]^2}{n}}$$

Where: σ =   the **Population Standard Deviation** for the entire distribution, population, or data set of "**n**" different values of "x_i";

σ^2 =   the **Population Variance** for the entire distribution, population, or data set of "**n**" different values of "x_i";

x_i =   is the value of the "**ith**" of "**n**" members of the overall distribution, population, or data set under consideration;

n   =   the number of members in the overall distribution, population, or data set under consideration; and

$$\mu = \overline{x} \quad = \quad \text{the Mean of the distribution, population, or data set.}$$

Equation 8.11

The following Equation, 8.11, defines the **Sample Coefficient of Variation** or **Relative Standard Deviation**, which is the first measure of the specific dispersion of all the data in any population, distribution, or data set being considered. This expression is shown in two identical forms below:

$$CV_{sample} \; = \; \frac{s}{\mu} \; = \; \frac{s}{\overline{x}}$$

Where: CV_{sample} = the **Sample Coefficient of Variation** for any population, distribution, or data set of "**n**" different values of "x_i";

s = the Sample Standard Deviation for the entire distribution, population, or data set of "**n**" different values of "x_i"; and

$\mu = \overline{x}$ = the Mean of the distribution, population, or data set.

Equation 8.12

The following Equation, 8.12, defines the **Population Coefficient of Variation**, which is the second measure of the specific dispersion of all the data in any population, distribution, or data set being considered. Proceeding logically from the previous relationship — i.e., Equation 8.11 — this one has been provided below in two useful formats:

$$CV_{population} \; = \; \frac{\sigma}{\mu} \; = \; \frac{\sigma}{\overline{x}}$$

Where: $CV_{population}$ = the **Population Coefficient of Variation** for the population, distribution, or data set of "**n**" different values of "x_i";

σ = the Population Standard Deviation for the entire distribution, population, or data set of "**n**" different values of "x_i";

$\mu = \overline{x}$ = the Mean of the distribution, population, or data set.

Appendix A
Conversion Factors
Alphabetical Listing

1 atm (atmosphere) =
    1.013 bars

    10.133 newtons/cm² (newtons/square centimeter)

    33.90 ft. of H_2O (feet of water)

    101.325 kp (kilopascals)

    1,013.25 mb (millibars)

    14.70 psia (pounds/square inch – absolute)

    760 torr

    760 mm Hg (millimeters of mercury)

1 bar =
    0.987 atm (atmospheres)

    1×10^6 dynes/cm² (dynes/square centimeter)

    33.45 ft. of H_2O (feet of water)

    1×10^5 pascals [nt/m²] (newtons/square meter)

    750.06 torr

    750.06 mm Hg (millimeters of mercury)

1 Bq (becquerel) =
    1 radioactive disintegration/second

    2.7×10^{-11} Ci (curie)

    2.7×10^{-8} mCi (millicurie)

1 BTU (British Thermal Unit) =
    252 cal (calories)

    1,055.06 j (joules)

    10.41 liter-atmospheres

    0.293 watt-hours

1 cal (calorie) =
    3.97×10^{-3} BTUs (British Thermal Units)

    4.18 j (joules)

    0.0413 liter-atmospheres

    1.163×10^{-3} watt-hours

1 cm (centimeter) =   0.0328 ft (feet)

0.394 in (inches)

10,000 microns (micrometers)

100,000,000 Å = 10^8 Å (Ångstroms)

1 cc (cubic centimeter) = 3.53×10^{-5} ft^3 (cubic feet)

0.061 in^3 (cubic inches)

2.64×10^{-4} gal (gallons)

0.001 ℓ (liters)

1.00 ml (milliliters)

1 ft^3 (cubic foot) =   28,317 cc (cubic centimeters)

1,728 in^3 (cubic inches)

0.0283 m^3 (cubic meters)

7.48 gal (gallons)

28.32 ℓ (liters)

29.92 qts (quarts)

1 in^3 (cubic inch) =   16.39 cc (cubic centimeters)

16.39 ml (milliliters)

5.79×10^{-4} ft^3 (cubic feet)

1.64×10^{-5} m^3 (cubic meters)

4.33×10^{-3} gal (gallons)

0.0164 ℓ (liters)

0.55 fl oz (fluid ounces)

1 m^3 (cubic meter) =   1,000,000 cc = 10^6 cc (cubic centimeters)

35.31 ft^3 (cubic feet)

61,023 in^3 (cubic inches)

264.17 gal (gallons)

1,000 ℓ (liters)

1 yd^3 (cubic yard) =   201.97 gal (gallons)

764.55 ℓ (liters)

1 Ci (curie) = 3.7×10^{10} radioactive disintegrations/second

3.7×10^{10} Bq (becquerel)

1,000 mCi (millicurie)

$1\ day =$ 24 hrs (hours)

1,440 min (minutes)

86,400 sec (seconds)

0.143 weeks

2.738×10^{-3} yrs (years)

$1\ ^{\circ}C\ (expressed\ as\ an\ interval) =$ $1.8^{\circ}F = [9/5]^{\circ}F$ (degrees Fahrenheit)

$1.8^{\circ}R$ (degrees Rankine)

$1.0\ K$ (degrees Kelvin)

$^{\circ}C\ (degree\ Celsius) =$ $[(5/9)(^{\circ}F - 32^{\circ})]$

$1\ ^{\circ}F\ (expressed\ as\ an\ interval) =$ $0.556^{\circ}C = [5/9]^{\circ}C$ (degrees Celsius)

$1.0^{\circ}R$ (degrees Rankine)

$0.556\ K$ (degrees Kelvin)

$^{\circ}F\ (degree\ Fahrenheit) =$ $[(9/5)(^{\circ}C) + 32)^{\circ}]$

$1\ dyne =$ 1×10^{-5} nt (newtons)

$1\ ev\ (electron\ volt) =$ 1.602×10^{-12} ergs

1.602×10^{-19} j (joules)

$1\ erg =$ 1 dyne-centimeters

1×10^{-7} j (joules)

2.78×10^{-11} watt-hours

$1\ fps\ (feet/second) =$ 1.097 kmph (kilometers/hour)

0.305 mps (meters/second)

0.01136 mph (miles/hour)

$1\ ft\ (foot) =$ 30.48 cm (centimeters)

12 in (inches)

0.3048 m (meters)

1.65×10^{-4} nt (nautical miles)

1.89×10^{-4} mi (statute miles)

$1\ gal\ (gallon) =$ 3,785 cc (cubic centimeters)

0.134 ft^3 (cubic feet)

231 in^3 (cubic inches)

3.785 ℓ (liters)

1 gm (gram) = 0.001 kg (kilograms)
1,000 mg (milligrams)
1,000,000 ng = 10^6 ng (nanograms)
2.205×10^{-3} lbs (pounds)

1 gm/cc (grams/cubic centimeter) = 62.43 lbs/ft^3 (pounds/cubic foot)
0.0361 lbs/in^3 (pounds/cubic inch)
8.345 lbs/gal (pounds/gallon)

1 Gy (gray) = 1 j/kg (joules/kilogram)
100 rad
1 Sv (sievert) — unless modified through division by an appropriate factor, such as Q and/or N

1 hp (horsepower) = 745.7 j/sec (joules/sec)

1 hr (hour) = 0.0417 days
60 min (minutes)
3,600 sec (seconds)
5.95×10^{-3} weeks
1.14×10^{-4} yrs (years)

1 in (inch) = 2.54 cm (centimeters)
1,000 mils

1 inch of water = 1.86 mm Hg (millimeters of mercury)
249.09 pascals
0.0361 psi (lbs/in^2)

1 j (joule) = 9.48×10^{-4} BTUs (British Thermal Units)
0.239 cal (calories)
10,000,000 ergs = 1×10^7 ergs
9.87×10^{-3} liter-atmospheres
1.00 nt-m (newton-meters)

1 kcal (kilocalorie) = 3.97 BTUs (British Thermal Units)
1,000 cal (calories)
4,186.8 j (joules)

1 kg (kilogram) =	1,000 gms (grams)
	2.205 lbs (pounds)
1 km (kilometer) =	3,280.8 ft (feet)
	0.54 nt (nautical miles)
	0.6214 mi (statute miles)
1 kw (kilowatt) =	56.87 BTU/min (British Thermal Units/minute)
	1.341 hp (horsepower)
	1,000 j/sec (joules/sec)
1 kw-hr (kilowatt-hour) =	3,412.14 BTU (British Thermal Units)
	3.6×10^6 j (joules)
	859.8 kcal (kilocalories)
1 ℓ(liter) =	1,000 cc (cubic centimeters)
	1 dm^3 (cubic decimeters)
	0.0353 ft^3 (cubic feet)
	61.02 in^3 (cubic inches)
	0.264 gal (gallons)
	1,000 ml (milliliters)
	1.057 qts (quarts)
1 m (meter) =	1×10^{10} Å (Ångstroms)
	100 cm (centimeters)
	3.28 ft (feet)
	39.37 in (inches)
	1×10^{-3} km (kilometers)
	1,000 mm (millimeters)
	$1,000,000 \mu = 1 \times 10^6 \mu$ (micrometers)
	1×10^9 nm (nanometers)
1 mps (meters/second) =	196.9 fpm (feet/minute)
	3.6 kmph (kilometers/hour)
	2.237 mph (miles/hour)
1 mph (mile/hour) =	88 fpm (feet/minute)
	1.61 kmph (kilometers/hour)
	0.447 mps (meters/second)

1 kt (nautical mile) =   6,076.1 ft (feet)
1.852 km (kilometers)
1.15 mi (statute miles)
2,025.4 yds (yards)

1 mi (statute mile) =   5,280 ft (feet)
1.609 km (kilometers)
1,609.3 m (meters)
0.869 nt (nautical miles)
1,760 yds (yards)

1 mCi (millicurie) =   0.001 Ci (curie)
3.7×10^{10} radioactive disintegrations/second
3.7×10^{10} Bq (becquerel)

1 mm Hg (millimeter of mercury) = 1.316×10^{-3} atm (atmospheres)
0.535 in H_2O (inches of water)
1.33 mb (millibars)
133.32 pascals
1 torr
0.0193 psia (pounds/square inch – absolute)

1 min (minute) = 6.94×10^{-4} days
0.0167 hrs (hours)
60 sec (seconds)
9.92×10^{-5} weeks
1.90×10^{-6} yrs (years)

1 nt (newton) = 1×10^{5} dynes

1 nt-m (newton-meter) =   1.00 j (joules)
2.78×10^{-4} watt-hours

1 ppm (parts/million-volume) =   1.00 ml/m^3 (milliliters/cubic meter)

1 ppm[wt] (parts/million-weight) =   1.00 mg/kg (milligrams/kilogram)

1 pascal = 9.87×10^{-6} atm (atmospheres)
4.015×10^{-3} in H_2O (inches of water)
0.01 mb (millibars)
7.5×10^{-3} mm Hg (mm of mercury)

1 lbs (pound) =	453.59 gms (grams)
	16 oz (ounces)
1 lbs/ft³ (pounds/cubic foot) =	16.02 gms/l (grams/liter)
1 lbs/in³ (pounds/cubic inch) =	27.68 gms/cc (grams/cubic centimeter)
	1,728 lbs/ft³ (pounds/cubic foot)
1 psi (pounds/square inch) =	0.068 atm (atmospheres)
	27.67 in H_2O (inches of water)
	68.85 mb (millibars)
	51.71 mm Hg (millimeters of mercury)
	6,894.76 pascals
1 qt (quart) =	946.4 cc (cubic centimeters)
	57.75 in³ (cubic inches)
	0.946 ℓ (liters)
1 rad =	100 ergs/gm (ergs/gram)
	0.01 Gy (gray)
	1 rem — unless modified through division by an appropriate factor, such as Q and/or N
1 rem =	1 rad — unless modified through division by an appropriate factor, such as Q and/or N
1 Sv (sievert) =	1 Gy (gray) — unless modified through division by an appropriate factor, such as Q and/or N
1 cm² (square centimeter) =	1.076×10^{-3} ft² (square feet)
	0.155 in² (square inches)
	1×10^{-4} m² (square meters)
1 ft² (square foot) =	2.296×10^{-5} acres
	929.03 cm² (square centimeters)
	144 in² (square inches)
	0.0929 m² (square meters)

1 m² (square meter) =	10.76 ft² (square feet)
	1,550 in² (square inches)
1 mi² (square mile) =	640 acres
	2.79×10^7 ft² (square feet)
	2.59×10^6 m² (square meters)
1 torr =	1.33 mb (millibars)
1 watt =	3.41 BTU/hr (British Thermal Units/hour)
	1.341×10^{-3} hp (horsepower)
	1.00 j/sec (joules/second)
1 watt-hour =	3.412 BTUs (British Thermal Units)
	859.8 cal (calories)
	3,600 j (joules)
	35.53 liter-atmospheres
1 week =	7 days
	168 hrs (hours)
	10,080 min (minutes)
	6.048×10^5 sec (seconds)
	0.0192 yrs (years)
1 yr (year) =	365.25 days
	8,766 hrs (hours)
	5.26×10^5 min (minutes)
	3.16×10^7 sec (seconds)
	52.18 weeks

Appendix B
Conversion Factors Listing By Unit Category

Units of Length

1 cm (centimeter) =	0.0328 ft (feet)
	0.394 in (inches)
	10,000 microns (micrometers)
	100,000,000 Å = 10^8 Å (Ångstroms)
1 ft (foot) =	30.48 cm (centimeters)
	12 in (inches)
	0.3048 m (meters)
	1.65×10^{-4} nt (nautical miles)
	$1.89 \times 10 - 4$ mi (statute miles)
1 in (inch) =	2.54 cm (centimeters)
	1,000 mils
1 km (kilometer) =	3,280.8 ft (feet)
	0.54 nt (nautical miles)
	0.6214 mi (statute miles)
1 m (meter) =	1×10^{10} Å (Ångstroms)
	100 cm (centimeters)
	3.28 ft (feet)
	39.37 in (inches)
	1×10^{-3} km (kilometers)
	1,000 mm (millimeters)
	1,000,000 μ = 1×10^6 μ (micrometers)
	1×10^9 nm (nanometers)
1 kt (nautical mile) =	6,076.1 ft (feet)
	1.852 km (kilometers)
	1.15 mi (statute miles)
	2,025.4 yds (yards)

159

1 mi (statute mile) =   5,280 ft (feet)
1.609 km (kilometers)
1,609.3 m (meters)
0.869 nt (nautical miles)
1,760 yds (yards)

Units of Area

1 cm² (square centimeter) = 1.076×10^{-3} ft² (square feet)
0.155 in² (square inches)
1×10^{-4} m² (square meters)

1 ft² (square foot) = 2.296×10^{-5} acres
929.03 cm² (square centimeters)
144 in² (square inches)
0.0929 m2 (square meters)

1 m² (square meter) =   10.76 ft² (square feet)
1,550 in² (square inches)

1 mi² (square mile) =   640 acres
2.79×10^{7} ft² (square feet)
2.59×10^{6} m2 (square meters)

Units of Volume

1 cc (cubic centimeter) = 3.53×10^{-5} ft³ (cubic feet)
0.061 in³ (cubic inches)
2.64×10^{-4} gal (gallons)
0.001 ℓ (liters)
1.00 ml (milliliters)

1 ft³ (cubic foot) =   28,317 cc (cubic centimeters)
1,728 in³ (cubic inches)
0.0283 m³ (cubic meters)
7.48 gal (gallons)
28.32 ℓ (liters)
29.92 qts (quarts)

1 in³ (cubic inch) =   16.39 cc (cubic centimeters)
16.39 ml (milliliters)
5.79×10^{-4} ft³ (cubic feet)
1.64×10^{-5} m³ (cubic meters)

4.33×10^{-3} gal (gallons)

0.0164 *ℓ* (liters)

0.55 fl oz (fluid ounces)

1 m³ (cubic meter) = $1,000,000 \ cc = 10^6$ cc (cubic centimeters)

35.31 ft³ (cubic feet)

61,023 in³ (cubic inches)

264.17 gal (gallons)

1,000 *ℓ* (liters)

1 yd³ (cubic yard) =   201.97 gal (gallons)

764.55 *ℓ* (liters)

1 gal (gallon) =   3,785 cc (cubic centimeters)

0.134 ft³ (cubic feet)

231 in³ (cubic inches)

3.785 *ℓ* (liters)

1 ℓ (liter) =   1,000 cc (cubic centimeters)

1 dm³ (cubic decimeters)

0.0353 ft³ (cubic feet)

61.02 in³ (cubic inches)

0.264 gal (gallons)

1,000 ml (milliliters)

1.057 qts (quarts)

1 qt (quart) =   946.4 cc (cubic centimeters)

57.75 in³ (cubic inches)

0.946 *ℓ* (liters)

Units of Mass

1 gm (gram) =   0.001 kg (kilograms)

1,000 mg (milligrams)

$1,000,000 \ ng = 10^6$ ng (nanograms)

2.205×10^{-3} lbs (pounds)

1 kg (kilogram) =   1,000 gms (grams)

2.205 lbs (pounds)

1 lb (pound) =   453.59 gms (grams)

16 oz (ounces)

Units of Time

1 day =	24 hrs (hours)
	1,440 min (minutes)
	86,400 sec (seconds)
	0.143 weeks
	2.738×10^{-3} yrs (years)
1 hr (hour) =	0.0417 days
	60 min (minutes)
	3,600 sec (seconds)
	5.95×10^{-3} weeks
	1.14×10^{-4} yrs (years)
1 min (minute) =	6.94×10^{-4} days
	0.0167 hrs (hours)
	60 sec (seconds)
	9.92×10^{-5} weeks
	1.90×10^{-6} yrs (years)
1 week =	7 days
	168 hrs (hours)
	10,080 min (minutes)
	6.048×10^{5} sec (seconds)
	0.0192 yrs (years)
1 yr (year) =	365.25 days
	8,766 hrs (hours)
	5.26×10^{5} min (minutes)
	3.16×10^{7} sec (seconds)
	52.18 weeks

Units of the Measure of Temperature

°C (degree Celsius) =	$[(5/9)(°F - 32°)]$
1 °C (expressed as an interval) =	$1.8°F = [9/5]°F$ (degrees Fahrenheit)
	$1.8°R$ (degrees Rankine)
	1.0 K (degrees Kelvin)
°F (degree Fahrenheit) =	$[(9/5)(°C) + 32°]$

1 °F (expressed as an interval) = $0.556°C = [5/9]°C$ (degrees Celsius)

1.0°R (degrees Rankine)

0.556 K (degrees Kelvin)

Units of Force

1 dyne = $1 \times 10 -5$ nt (newtons)

1 nt (newton) = 1×10^5 dynes

Units of Work or Energy

1 BTU (British Thermal Unit) =   252 cal (calories)

1,055.06 j (joules)

10.41 liter-atmospheres

0.293 watt-hours

1 cal (calorie) = $3.97 \times 10 -3$ BTUs (British Thermal Units)

4.18 j (joules)

0.0413 liter-atmospheres

$1.163 \times 10 -3$ watt-hours

1 ev (electron volt) = $1.602 \times 10 -12$ ergs

$1.602 \times 10 -19$ j (joules)

1 erg =   1 dyne-centimeter

$1 \times 10 -7$ j (joules)

$2.78 \times 10 -11$ watt-hours

1 j (joule) = $9.48 \times 10 -4$ BTUs (British Thermal Units)

0.239 cal (calories)

10,000,000 ergs = $1 \times 10 7$ ergs

$9.87 \times 10 -3$ liter-atmospheres

1.00 nt-m (newton-meters)

1 kcal (kilocalorie) =   3.97 BTUs (British Thermal Units)

1,000 cal (calories)

4,186.8 j (joules)

1 kw-hr (kilowatt-hour) =	3,412.14 BTU (British Thermal Units)
	3.6×10^6 j (joules)
	859.8 kcal (kilocalories)
1 nt-m (newton-meter) =	1.00 j (joules)
	2.78×10^{-4} watt-hours
1 watt-hour =	3.412 BTUs (British Thermal Units)
	859.8 cal (calories)
	3,600 j (joules)
	35.53 liter-atmospheres

Units of Power

1 hp (horsepower) =	745.7 j/sec (joules/sec)
1 kw (kilowatt) =	56.87 BTU/min (British Thermal Units/minute)
	1.341 hp (horsepower)
	1,000 j/sec (joules/sec)
1 watt =	3.41 BTU/hr (British Thermal Units/hour)
	1.341×10^{-3} hp (horsepower)
	1.00 j/sec (joules/second)

Units of Pressure

1 atm (atmosphere) =	1.013 bars
	10.133 newtons/cm^2 (newtons/square centimeter)
	33.90 ft. of H$_2$O (feet of water)
	101.325 kp (kilopascals)
	1,013.25 mb (millibars)
	14.70 psia (pounds/square inch – absolute)
	760 torr
	760 mm Hg (millimeters of mercury)
1 bar =	0.987 atm (atmospheres)
	1×10^6 dynes/cm^2 (dynes/square centimeter)

33.45 ft. of H_2O (feet of water)

1×10^5 pascals (nt/m^2)
(newtons/square meter)

750.06 torr

750.06 mm Hg (millimeters of
mercury)

1 inch of water =   1.86 mm Hg (millimeters of
mercury)

249.09 pascals

0.0361 psi (lbs/in^2)

1 mm Hg (millimeter of mercury) = 1.316×10^{-3} atm (atmospheres)

0.535 in H_2O (inches of water)

1.33 mb (millibars)

133.32 pascals

1 torr

0.0193 psia (pounds/square inch —
absolute)

1 pascal = 9.87×10^{-6} atm (atmospheres)

4.015×10^{-3} in H_2O (inches of
water)

0.01 mb (millibars)

7.5×10^{-3} mm Hg (millimeters of
mercury)

1 psi (pounds/square inch) =   0.068 atm (atmospheres)

27.67 in H_2O (inches of water)

68.85 mb (millibars)

51.71 mm Hg (millimeters of
mercury)

6,894.76 pascals

1 torr =   1.33 mb (millibars)

Units of Velocity or Speed

1 fps (feet/second) =   1.097 kmph (kilometers/hour)

0.305 mps (meters/second)

0.01136 mph (miles/hour)

1 mps (meters/second) =	196.9 fpm (feet/minute)
	3.6 kmph (kilometers/hour)
	2.237 mph (miles/hour)
1 mph (mile/hour) =	88 fpm (feet/minute)
	1.61 kmph (kilometers/hour)
	0.447 mps (meters/second)

Units of Density

1 gm/cc (grams/cubic centimeter) =	62.43 lbs/ft^3 (pounds/cubic foot)
	0.0361 lbs/in^3 (pounds/cubic inch)
	8.345 lbs/gal (pounds/gallon)
1 lbs/ft^3 (pounds/cubic foot) =	16.02 gms/ℓ (grams/liter)
1 lbs/in^3 (pounds/cubic inch) =	27.68 gms/cc (grams/cubic centimeter)
	1,728 lbs/ft^3 (pounds/cubic foot)

Units of Concentration

| *1 ppm (parts/million-volume) =* | 1.00 ml/m^3 (milliliters/cubic meter) |
| *1 ppm(wt) (parts/million-weight) =* | 1.00 mg/kg (milligrams/kilogram) |

Radiation and Dose Related Units

1 Bq (becquerel) =	1 radioactive disintegration/second
	2.7×10^{-11} Ci (curie)
	2.7×10^{-8} mCi (millicurie)
1 Ci (curie) =	3.7×10^{10} radioactive disintegrations/second
	3.7×10^{10} Bq (becquerel)
	1,000 mCi (millicurie)
1 Gy (gray) =	1 j/kg (joule/kilogram)
	100 rad
	1 Sv (sievert) — unless modified through division by an appropriate factor, such as Q and/or N

1 mCi (millicurie) =	0.001 Ci (curie)
	3.7×10^{10} radioactive disintegrations/second
	3.7×10^{10} Bq (becquerel)
1 rad =	100 ergs/gm (ergs/gm)
	0.01 Gy (gray)
	1 rem — unless modified through division by an appropriate factor, such as Q and/or N
1 rem =	1 rad — (unless modified through division by an appropriate factor, such as Q and/or N)
1 Sv (sievert) =	1 Gy (gray) — unless modified through division by an appropriate factor, such as Q and/or N

Appendix C
The Atmosphere

Components of the Atmosphere

Because virtually all of the measurements that are made by professionals working in the fields of occupational safety and health, industrial hygiene, and/or the environment are completed in the earth's atmosphere, it is important to understand the nature of the negative impacts that this atmospheric matrix might have on some of these measurements (i.e., negative impacts = factors, circumstances, atmospheric components, etc. that will tend to cause a particular measurement to be incorrect). Certainly many of the measurements that are made in the atmosphere are totally unaffected by any of the elements or compounds that make it up; however, there are quite a few for which this is not true. To understand the relationships between the atmospheric matrix and the measurements that are made in it, we must know its major components, as well as the concentration range of each.

The most common way to tabulate these components is to consider *dry air* — i.e., air in which there is absolutely no water vapor. Clearly, nowhere on earth is there ever any air mass that is completely free of water vapor. In fact, for most situations, water vapor will exist at a concentration that would place it in the third highest position, trailing only nitrogen and oxygen. The range of the ambient concentrations of water vapor varies over more than 2.5 orders of magnitude — i.e., ~100 ppm(vol) at points in the Antarctic in winter, to 35,000+ in an equatorial rain forest. Because water vapor can have such a wide concentration range, as stated above, the most common way to list the components that make up the air is to consider them on a *dry*, or water vapor free, basis. Table 1 lists the seventeen most common atmospheric components in the order of decreasing concentrations, again on a water vapor free or *dry* basis.

Of the components listed on the previous page, seven have been shown both in terms of their average atmospheric concentration, as well as the range of ambient concentrations common for each. For the remaining ten components, the generally accepted, ambient, dry basis concentration value has been provided.

For the seven components having both an average concentration and a concentration range, there are a variety of activities, situations, and naturally occurring circumstances that are responsible for producing the observed concentration variations. Included among these factors are such things as: industrial production, changes in rain forest acreage, volcanic eruptions, etc. Clearly any of these factors can and do occur in a wide variety of locations on the earth, at any time of the day or night, etc. For the seven variable concentration components, some of the specific factors,

Table 1.   Most common atmospheric components

No.	Component	Concentration – in ppm(vol)
1.	Nitrogen	780,840
2.	Oxygen	209,459
3.	Argon	9,303
4.	Carbon Dioxide	average = 370; range: 350 to 580
5.	Neon	18.2
6.	Helium	5.2
7.	Methane	average = 2.8; range: 2.5 to 4.5
8.	Krypton	1.1
9.	Hydrogen	0.4
10.	Nitrous Oxide	average = 0.3; range: 0.3 to 0.6
11.	Sulfur Dioxide	average = 0.2; range: 0.2 to 0.3
12.	Xenon	0.07
13.	Ozone	average = 0.02; range: 0.02 to 0.8
14.	Nitrogen Dioxide	average = 0.02; range: 0.01 to 0.7
15.	Carbon Monoxide	average = 0.01; range: 0.01 to 0.7
16.	Iodine	< 0.01
17.	Ammonia	< 0.01
	Summation	*1,000,000.32*

activities, situations, and naturally occurring circumstances that are responsible for producing these specific concentration changes:

Carbon Dioxide   combustion of organic fuels — i.e., coal, gasoline, etc.

Methane   decay of organic matter in the soil, mammalian flatulence, etc.

Nitrous Oxide   industrial production, electrical discharges — i.e., lightning

Sulfur Dioxide   volcanic activity, combustion of high sulfur coal

Ozone   solar activity, industrial activity, electrical discharges

Nitrogen Dioxide   combustion of organic fuels in the presence of excess air

Carbon Monoxide   incomplete combustion of organic fuels

There are several common atmospheric concentration measurements that experience interferences — some positive, some negative — from one or more of the seventeen components of the air. A listing of some of the frequently made ambient concentration measurements that experience interferences from the various atmospheric components is shown in Table 2.

In this tabulation, both the specific measurement and its most likely measurement technology have been listed against the interfering component and the direction of the anticipated interference — with this interference direction shown following the component in brackets as "+", for positive or additive interferences, and "−" for negative or subtractive interferences.

Table 2.   Frequently made ambient concentration measurements

No.	Measurement	Monitoring Technology	Interfering Component
1.	Ambient Hydrocarbons	Flame Ionization Detectors	Methane [+]
2.	Organic Vapors	Photo Ionization Detectors	Methane [+ or −]
3.	Mercury Vapors	UV Photometry	Ozone [+]
4.	SF_6 Tracer Gas	ECD Chromatography	Oxygen [++]

This listing is by no means complete; however, it does show four of the common ambient measurements that are impacted by various atmospheric components. The first two of these measurements are associated with routine air pollution monitoring. The measurement of ambient mercury vapor concentrations is a common industrial hygiene measurement made in any situation where mercury is used — i.e., mining, high voltage rectification, etc. The measurement of the ambient levels of the tracer, sulfur hexafluoride [SF_6], is virtually always accomplished by using a gas chromatograph equipped with an Electron Capture Detector [ECD]. This methodology is used because of the extremely low concentration level of SF_6 (i.e., ppb and even ppt by volume) that is common in atmospheric tracer work.

On balance, as a result of the earth's gravitational field, one would expect that the heavier molecules or atoms from the Table 1 listing would tend to concentrate at lower elevations, while the reverse would be true for the lighter components. This seems logical because in indoor situations, it has long been recognized that heavy molecules such as carbon dioxide tend to concentrate at the floor level. On this basis, the heaviest and the lightest three or four components listed on the following page, as Table 3 (Heavy Components) and Table 4 (Light Components), logically would be expected to exhibit this sort of behavior — i.e., the heavier members, down; and lighter ones, up.

Table 3.   Heavy Components

No.	Component	Atomic or Molecular Weights
12.	Xenon	131.3 amu
16.	Iodine	126.9 amu
8.	Krypton	83.8 amu
11.	Sulfur Dioxide	64.0 amu

Note: *the "effective" molecular weight of air is 28.964 amu; thus each of the following five listed atmospheric components — including* water, *which heretofore has not been discussed — is "effectively" lighter than the air mass in which each is a component. Again, the basic laws of physics would seem to apply, with the result being a situation where each of these components should experience a buoyant force, from the "heavier" air matrix in which they exist, and therefore tend to lift or loft up to steadily higher altitudes. That this effect does not occur results from the highly efficient meteorological mixing of the atmosphere that occurs constantly at all points and altitudes on the Earth.*

Table 4.   Light Components

No.	Component	Atomic or Molecular Weights
9.	Hydrogen	2.0 amu
6.	Helium	4.0 amu
7.	Methane	16.0 amu
17.	Ammonia	17.0 amu
—	Water	18.0 amu

Although the earth's atmosphere is very well mixed, the volume-based concentrations listed in Table 1 are accurate for *all* terrestrial altitudes and/or barometric pressures — i.e., from the level of the Dead Sea to the top of Mount Everest. Of course, if the concentrations of these components were to have been expressed on a mass per unit volume basis — i.e., in units such as mg/m^3 — then there would have been a wide variation in the listed concentrations, heavily dependent on the altitude where the measurements were made.

As an example, the ambient concentration of Argon, expressed both in volumetric- and mass-based units, at five identifiable locations on the Earth (including the altitude of each location) is given in Table 5.

Table 5.   Ambient concentration of Argon

Location	Altitude relative to Mean Sea Level	Ambient Concentration ppm(vol)	mg/m^3
Dead Sea, Israel	− 396 meters	9,303 ppm	17,377 mg/m³
At Sea Level	0 meters	9,303 ppm	16,581 mg/m³
Denver, CO,	+ 1,609 meters	9,303 ppm	13,623 mg/m³
Lake Titicaca, Peru	+ 3,806 meters	9,303 ppm	10,343 mg/m³
Atop Mt. Everest, Nepal	+ 8,848 meters	9,303 ppm	5,065 mg/m3

Atmospheric Humidity —
Ambient Water Vapor Concentrations

The concentration of water vapor in the atmosphere can and does vary over a wide range of values, as stated earlier. Because of these variations in concentration, water vapor is usually omitted from any tabulation of the component materials that make up the air — i.e., see Table 1.

Whenever one thinks about the ambient water vapor concentration, the perspective is of a measurement expressed in one of the two common "water vapor specific" sets of concentration units, units that *never* are applied to the measurement of the concentration of any other component of the air. These two widely used units of water vapor concentration are:

1. the Relative Humidity at some average Ambient Temperature, and
2. the Dew Point.

Rarely, if ever, does one hear about the water vapor concentration in what might be a more useful format, namely, in some unit of its "absolute humidity", rather than its "relative humidity" cousin. It is important here to note the specific differences between these concentration units, and the most effective way to accomplish this is to reproduce the definitions of each *Note:* These definitions also are provided in Chapter 4 — *Ventilation.*

Absolute Humidity   The specific amount of vaporous water in the atmosphere, measured on a basis of its mass per unit volume. Absolute humidity is most frequently expressed in units of lbs/ft^3 or mg/m^3.

Relative Humidity   The *ratio* of the actual or measured vapor pressure of water vapor in an air mass of some known temperature, to the saturated vapor pressure of pure water, at the same temperature as that of the air mass. Relative Humidity can be quantified as the value of the following ratio [determined at some specific known ambient temperature]. *Note:* The ratio must be multiplied by 100 to convert it into a percent.

$$RH = 100\left[\frac{\text{Measured Absolute Humidity, in mg / m}^{\cdot}}{\text{Maximum Possible Absolute Humidity, in mg / m}^{\cdot}}\right]$$

Dew Point   The Dew Point of an air mass is also a measure of relative humidity. It is the temperature at which the water vapor present in that air mass would start to condense — or in the context of the term, Dew Point, it is the temperature at which that water vapor would turn to "dew". It can be determined experimentally simply by cooling a mass of air until "dew", or condensate, forms; the temperature at which this event occurs is defined to be the Dew Point of the air mass that had been cooled.

Let us next consider two tabulations of the actual concentration of water in ambient air (i.e., the absolute humidity) vs. various specific values of relative humidity. In the first of these two tabulations, Table 6, the water vapor concentrations will be volumetric, namely, ppm(vol): (1) at five specific relative humidities, and (2) at the Ambient Temperature at which each of the five relative humidities was determined.

The water vapor concentrations in Table 6 at the 100% relative humidity level are — by definition — the concentrations that would exist in any atmosphere at its Dew Point. Condensation always forms at the Dew Point because the air is saturated with water vapor under those conditions — i.e., its relative humidity equals 100%. Thus, whenever one wants the ambient water vapor volume-based concentration at Sea Level for some identifiable dew point, this concentration can either be (1) taken directly from the last column of Table 6, or (2) determined by linear interpolation between adjacent pairs of values taken from this column for those situations in which the temperature involved is not among those tabulated.

In every listing in Table 6, it has been assumed that the prevailing ambient barometric pressure is 760 mm Hg, or 1.0 atmosphere — i.e., the barometric pressure at Sea Level. At different (notably lesser) barometric pressures, the values in the last two columns of Table 6 will likely be incorrect; however the values in all the other columns will still be accurate.

Finally, if it is ever necessary to determine a Sea Level volume-based water vapor concentration at some untabulated ambient temperature and/or relative humidity (i.e., the water vapor concentration at a temperature of 13.2°C, and a relative humidity of 67%), a simple horizontal and/or vertical linear interpolation from the data in Table 6 will provide a reasonably accurate result, certainly to within ±1 or 2% of the true value.

For reference, the water vapor concentration under the example conditions listed in the final paragraph on the previous page — obtained by a combination of horizontal and vertical linear interpolation — would be 9,997.3 ppm(vol).

Table 7 is analogous to Table 6. For this second tabulation, the water vapor concentrations have been listed in mass-based units of concentration, namely, mg/m^3. For most situations that require or involve knowing the ambient water vapor concentration on a true or absolute humidity basis, the mass-based concentration units in Table 7 will be the most useful reference.

As was true for Table 6, the water vapor concentrations tabulated in Table 7C at the 100% relative humidity level are again — by definition — the ambient concentrations, or absolute humidity levels, that would exist in the listed atmosphere at its Dew Point. Thus, whenever one wants the ambient Sea Level water vapor concentration in mg/m^3 at some identifiable Dew Point, this concentration can either be: (1) taken directly from the last column of Table 7, or (2) determined by linear interpolation between adjacent pairs of values taken from this column.

Table 6.   Concentration of water in ambient air vs. relative humidity

Temps.		Volume-based Water Vapor Concentrations, in ppm(vol) at Various Levels of Relative Humidity and at 760 mm Hg					
°C	°F	0% RH	20% RH	40% RH	60% RH	80% RH	100% RH
−20	−4.0	0	214	428	643	857	1,071
−18	−0.4	0	258	517	775	1,034	1,292
−16	3.2	0	315	631	946	1,261	1,576
−14	6.8	0	384	768	1,153	1,537	1,921
−12	10.4	0	466	932	1,397	1,863	2,329
−10	14.0	0	560	1,120	1,680	2,240	2,800
−8	17.6	0	666	1,333	1,999	2,665	3,332
−6	21.2	0	785	1,571	2,356	3,141	3,926
−4	24.8	0	917	1,833	2,750	3,666	4,853
−2	28.4	0	1,061	2,121	3,182	4,242	5,303
0	32.0	0	1,237	2,474	3,710	4,947	6,184
2	35.6	0	1,447	2,895	4,342	5,790	7,237
4	39.2	0	1,658	3,316	4,973	6,631	8,289
6	42.8	0	1,895	3,790	5,684	7,579	9,474
8	46.4	0	2,158	4,316	6,473	8,631	10,789
10	50.0	0	2,447	4,895	7,342	9,790	12,237
12	53.6	0	2,763	5,526	8,290	11,053	13,816
14	57.2	0	3,132	6,263	9,395	12,526	15,658
16	60.8	0	3,553	7,105	10,658	14,210	17,763
18	64.4	0	4,026	8,053	12,079	16,106	20,132
20	68.0	0	4,553	9,105	13,658	18,210	22,763
21.1	70.0	0	4,895	9,789	14,684	19,579	24,474
22	71.6	0	5,158	10,316	15,473	20,631	25,789
23.9	75.0	0	5,804	11,608	17,413	23,217	29,021
24	75.2	0	5,842	11,684	17,527	23,369	29,211
26	78.8	0	6,605	13,210	19,816	26,421	33,026
28	82.4	0	7,447	14,895	22,342	29,790	37,237
30	86.0	0	8,368	16,737	25,105	33,474	41,842
32	89.6	0	9,421	18,842	28,263	37,684	47,105
34	93.2	0	10,553	21,105	31,658	42,210	52,763
36	96.8	0	11,816	23,632	35,447	47,263	59,079
38	100.4	0	13,184	26,368	39,553	52,737	65,921
40	104.0	0	14,684	29,368	44,053	58,737	73,421
42	107.6	0	16,316	32,632	48,947	65,263	81,579
46	111.2	0	18,079	36,158	54,237	72,316	90,395
48	114.8	0	20,000	40,000	60,000	80,000	100,000

As was true for Table 6, for every listing in Table 7 it has been assumed that the prevailing ambient barometric pressure is 760 mm Hg, or 1.0 atmosphere, the barometric pressure at Sea Level. At different barometric pressures, the values in *every column* of Table 7 will be *incorrect*. For each mass-based concentration shown in this table, the listed value will be correct, ±1%, at the temperature tabulated in the first two columns, and at the 1.0 atmosphere or 760 mm Hg barometric pressure.

Finally, if it is ever necessary to determine a mass-based absolute water vapor concentration at Sea Level for some untabulated ambient temperature and/or relative humidity [i.e., the water vapor concentration at a temperature of 62°F, and a relative humidity of 45%], a simple horizontal and/or vertical linear interpolation from the data in Table 7 will provide a reasonably accurate result, certainly to within ±1 to 2% of the true value.

For reference, the water vapor concentration under the example conditions listed immediately above in the previous paragraph — obtained by a combination of horizontal and vertical linear interpolation — would be 6,317.9 mg/m³.

In the event that it is ever necessary to determine an absolute, mass-based water vapor concentration at (1) an unlisted ambient temperature, (2) an unlisted relative humidity, and (3) a barometric pressure *other than* 1.0 atmosphere or 760 mm Hg, the procedure required to accomplish this is a bit more complicated. Basically the following two step process must be followed. The first step involves using a simple horizontal and/or vertical linear interpolation from the data in Table 6 — **not** Table 7 — in order to obtain the *volume-based* water vapor concentration at the required ambient temperature and relative humidity. *Note:* This volume-based concentration is applicable at a barometric pressure of 760 mm Hg or 1.0 atmosphere. Once this value has been determined, the second step requires the use of Equation **3-9,** Chapter 3, which will then provide the mass-based water vapor concentration under *any* prevailing ambient conditions. The required Equation is:

$$C_{mass} = \frac{P[MW]}{RT}[C_{volume}]$$

For determinations involving water, using as a value for R, the Universal Gas Constant,

$$62.36 \, ^{(liter)(mm\,Hg)}\!/_{(K)(mole)},$$

and since water has a molecular weight of 18.0153 amu, we get:

$$C_{mass} = [0.2886]\frac{P}{T}[C_{volume}]$$

This overall two step process, which obviously uses this specific relationship, will give the required mass-based absolute water vapor concentration at *any ambient temperature, barometric pressure, and level of relative humidity.*

Table 7.   This table is analogous to Table 6.

**Mass-Based Water Vapor Concentrations, in mg/m³ at Various
Temps.   Levels of Relative Humidity, and at 760 mm Hg**

°C	°F	0% RH	20% RH	40% RH	60% RH	80% RH	100% RH
− 20	− 4.0	0	185.8	371.5	557.3	743.1	928.8
− 18	− 0.4	0	222.3	444.7	667.0	889.4	1,111.7
− 16	3.2	0	269.1	538.2	807.3	1,076.4	1,345.6
− 14	6.8	0	325.5	651.0	976.5	1,302.0	1,627.5
− 12	10.4	0	391.6	783.2	1,174.8	1,566.4	1,958.0
− 10	14.0	0	467.2	934.4	1,401.6	1,868.9	2,336.1
− 8	17.6	0	551.8	1,103.6	1,655.4	2,207.2	2,759.0
− 6	21.2	0	645.3	1,290.6	1,935.9	2,581.2	3,226.5
− 4	24.8	0	791.7	1,583.5	2,375.2	3,166.9	3,958.7
− 2	28.4	0	858.8	1,717.5	2,576.3	3,435.1	4,293.8
0	32.0	0	993.4	1,986.8	2,979.4	3,972.8	4,966.2
2	35.6	0	1,153.9	2,307.8	3,461.8	4,615.7	5,769.6
4	39.2	0	1,312.1	2,624.2	3,936.3	5,248.4	6,560.6
6	42.8	0	1,489.0	2,977.9	4,466.9	5,955.8	7,444.8
8	46.4	0	1,683.6	3,367.1	5,050.7	6,734.2	8,417.8
10	50.0	0	1,896.0	3,792.0	5,688.1	7,584.1	9,480.1
12	53.6	0	2,125.7	4,251.3	6,377.0	8,502.6	10,628.3
14	57.2	0	2,392.3	4,784.6	7,176.8	9,569.1	11,961.4
16	60.8	0	2,695.1	5,390.2	8,085.4	10,780.5	13,475.6
18	64.4	0	3,033.6	6,067.2	9,100.7	12,134.3	15,167.9
20	68.0	0	3,406.6	6,813.3	10,219.9	13,626.6	17,033.2
21.1	70.0	0	3,649.0	7,298.0	10,947.0	14,596.0	18,245.0
22	71.6	0	3,833.3	7,666.7	11,500.0	15,333.4	19,166.7
23.9	75.0	0	4,289.9	8,579.8	12,869.7	17,159.6	21,449.1
24	75.2	0	4,312.8	8,625.6	12,938.3	17,251.1	21,563.9
26	78.8	0	4,843.4	9,686.8	14,530.3	19,373.7	24,217.1
28	82.4	0	5,424.7	10,849.4	16,274.2	21,698.9	27,123.6
30	86.0	0	6,055.4	12,110.8	18,166.1	24,221.5	30,276.9
32	89.6	0	6,772.4	13,544.7	20,317.1	27,089.4	33,861.8
34	93.2	0	7,536.4	15,072.8	22,609.3	30,145.7	37,682.1
36	96.8	0	8,384.0	16,768.0	25,159.9	33,535.9	41,919.9
38	100.4	0	9,294.8	18,589.6	27,884.4	37,179.2	46,474.0
40	104.0	0	10,286.2	20,572.4	30,858.5	41,144.7	51,430.9
42	107.6	0	11,356.6	22,713.2	34,069.7	45,426.3	56,782.9
46	111.2	0	12,426.1	24,852.3	37,278.4	49,704.6	62,130.7
48	114.8	0	13,660.9	27,321.8	40,982.6	54,643.5	68,304.4

Water vapor can cause problems for a variety of atmospheric measurements, mostly those designed to determine the ambient concentration of some chemical or particulate of interest.

Water vapor concentrations by themselves — unless they are extremely high [i.e., > 25,000 ppm(vol) or > 18,000 mg/m^3] — usually will not be the source of too many problems; the principal exception to this rule is the quantification, by particle diameter, of the size distributions of certain dusts. For this specific analytical category, high ambient water vapor concentration levels will have a tendency to cause various smaller particles to agglomerate together, thereby forming steadily larger sized particles. Such a situation will obviously skew the size distribution results, shifting the entire population upward as expressed in effective diameters, producing what might appear to be a less hazardous distribution [i.e., particles having the smallest diameters — namely, the respirable, the alveolic, and/or the thoracic fractions — are removed from the population simply as a result of being agglomerated together, thereby becoming particles that are larger, and correspondingly, less hazardous].

For determinations of the ambient concentration levels of certain vapors, water vapor concentration changes are the source of numerous problems. In this area, the measurements that are made must almost always be compensated for by the level of water vapor in the ambient air. Such compensations are usually fairly easy. The problems arise because of the manner in which ambient water vapor concentrations can vary both: (1) over short time intervals, and (2) from point-to-point.

Changes in water vapor concentration levels over short time intervals can occur indoors as a result of a variety of factors, among which are:

1. the reduction (or the increase, in certain situations) of indoor humidity levels produced by the operation of air conditioning equipment;
2. the equilibration of the indoor and outdoor humidity levels — to the extent that these levels differ — by the opening of windows, etc.; and
3. the operation of an evaporative cooler (i.e., a swamp cooler) during the summer months in desert areas having relatively dry climates.

This listing is not complete. However, it does show some of the common ways in which short time interval very large changes in indoor humidity levels can be produced. As to outdoor situations where humidity levels experience major shifts over a short time period, the most obvious circumstance that could cause such an event would be the passage of a weather front. In any of these man or nature caused situations, humidity changes of 30–50% can and do occur — i.e., from an 80% RH level @ 75°F [~23,217 ppm(vol) or ~17,139 mg/m^3] to a 50% RH level @ 68°F [~11,382 ppm(vol) or ~8 8,517 mg/m^3]. Such a shift would be a change of approximately 50%. Short time interval humidity shifts of these magnitudes can be produced by efficient air conditioning systems, swamp coolers, or weather fronts.

Point-to-point changes in the humidity level occur principally indoors. Typically, an individual would likely experience large changes in these levels if he were to pass from an area where the temperature and humidity had been set for optimum human comfort to an area where these parameters were established in order to provide perfect operating conditions for a computer or other sensitive machinery, equipment, or system. It always will be the large scale humidity changes — like those listed above — that will likely prove to have been responsible for producing problems for an individual who must make any of a variety of ambient measurements.

Examples of some but not all of the specific types of measurements that might or might not be affected by *increases* in humidity, along with the measurement technology most likely to have been used for each determination, are tabulated in Table 8.

Table 8.   Increasing Humidity Levels

No.	Measurement	Monitoring Technology	Direction of Interference
1	Organic Vapors	IR Spectrophotometry	plus
2	Organic Vapors	NDIR	mostly plus
3	Organic Vapors	Photo Ionization Detectors	minus
4	Organic Vapors	Flame Ionization Detectors	none
5	Any Measurable Vapor	Electrochemical Detectors	some plus, some minus
6	Any Measurable Vapor	Solid State Detectors	some plus, some minus
7	Combustible Gases	Catalytic Bead Detectors	none
8	Oxygen	Electrochemical Detectors	minus

This tabulation is by no means complete; however, it does identify some of the most important measurements that can be affected by increases in humidity.

Examples of some of the specific types of measurements that might or might not be affected by *decreases* in humidity, along with the measurement technology most likely to have been used for each determination, are tabulated in Table 9.

Table 9.   Decreasing Humidity Levels

No.	Measurement	Monitoring Technology	Direction of Interference
1	Organic Vapors	IR Spectrophotometry	minus
2	Organic Vapors	NDIR	none
3	Organic Vapors	Photo Ionization Detectors	plus
4	Organic Vapors	Flame Ionization Detectors	none
5	Any Measurable Vapor	Electrochemical Detectors	some minus, some plus
6	Any Measurable Vapor	Solid State Detectors	some minus, some plus
7	Combustible Gases	Catalytic Bead Detectors	none
8	Oxygen	Electrochemical Detectors	plus

For virtually all of the measurements that experience problems arising from changes in the ambient water vapor concentration, the mechanism of the problem — expressed in the form of a sequence of events — will be as follows:

1. The analyzer involved is zeroed, or zero checked, at a location that is known to be free of whatever analyte is to be quantified. Care is not exercised in this process to note the prevailing ambient humidity level at the location where this process is completed.

2. The individual doing the analysis then proceeds to the location where the measurements are to be made, an area where the humidity level differs from that at the location where the analytical system was zeroed, or zero checked. No notice of this change in humidity level is made by the analyst.

3. Measurements are then made in the desired location. If the direction of the interference is negative (minus), then the individual is fairly likely to recognize the problem, since he or she may see the analytical result appear as a negative concentration, something that cannot possibly exist. If the direction of the interference is positive (plus), it is unlikely that the analyst will be able to determine that his or her readings are in error on the high side. The readings will simply appear to be indications of possibly higher than expected concentrations of whatever analyte is being measured.

It is interesting to note that in virtually all of the problem areas identified in either Tables 8 or 9, the magnitude of the interference produced by changes in the humidity will turn out to have been only a very small percentage of the absolute change in the water vapor concentration. The problem is simply that the magnitude of the absolute changes in the water vapor concentration *can be extremely large* — i.e., ±5,000 ppm(vol), or even more.

Consider an example. Assume that the magnitude of the error caused by a change in the ambient water vapor concentration was determined — for some specific analysis — to be only ±0.2% of the actual change in water vapor concentration. In most cases such an error, particularly expressed as a percentage, would appear to be unimportant and could be ignored. The actual error produced for the example suggested on the previous page, namely, one produced by a change in the absolute humidity level of, say, ±5,000 ppm(vol) — calculated as ±0.2% of this change in the water vapor concentration level — would be in the range of ±10 ppm(vol). If the measurement being made was for the ambient concentration acetone, for which the OSHA PEL-TWA is 1,000 ppm(vol), an error of ±10 ppm(vol) would, in actual fact, be sufficiently unimportant that it could be ignored. In contrast, if the ambient concentration level of formaldehyde [OSHA PEL-TWA = 0.75 ppm(vol), and OSHA PEL-STEL = 2.0 ppm(vol)] was being sought, an error of ±10 ppm(vol) would be totally unacceptable.

The result of all this is that considerable attention must be paid to the prevailing humidity levels whenever the goal is the accurate determination of the ambient concentration of some important volatile compound or chemical. Very frequently, as stated earlier, a potential for humidity related problems does exist — problems that are both a function of analyte involved and the analytical method being used to make the determination.

On rare occasions, it may be necessary to know the make-up — on a volume-based concentration basis — of an air mass on an actual, or *nondry*, basis. In such a situation, the accurate composition of air mass must be calculated. The true atmospheric composition will clearly differ from the tabulated *dry* basis concentrations shown in Table 1 since the air mass is now no longer *dry*. The determination of the actual make-up of the air mass of interest will require adjustments to each of the volume-based concentrations of the "normal" *dry* basis components, adjustments that are obviously required in order to compensate for the amount of water vapor present. In essence, the air mass must be recognized as being no longer *dry*, or water vapor free.

To make the required corrections, the following formula should be used to determine the factor, f_{volume}, by which each of the *dry* volume-based concentrations shown in Table 1 on must be multiplied in order to produce the true wet-basis concentrations for each of these components.

$$f_{volume} = \left[1 - \frac{\text{Water Vapor Concentration, in ppm(vol)}}{10^6} \right]$$

Thus, if we are in a situation where the relative humidity is 80% at 70°F, and the water vapor concentration — see Table 6 — is 19,579 ppm(vol), we can determine the required multiplication factor as follows:

$$f_{volume} = \left[1 - \frac{19,579}{10^6} \right] = [1 - 0.0196] = 0.9804$$

Applying this factor to the seven highest concentration members of a *dry* air mass, as shown in Table 1, we would obtain the true composition of this atmosphere, on a wet-basis, as in Table 10.

Table 10.   Actual Composition of Air (wet-basis)

No.	Component	True Concentration – in ppm(vol)
1	Nitrogen	765,552
2	Oxygen	205,258
–	water	19,579
3	Argon	9,121
4	Carbon Dioxide	average = 363; range: 343 to 569
5	Neon	17.8
6	Helium	5.1
7	Methane	Average = 2.7; range: 2.5 to 4.4

Note: For this air mass, water vapor is the member with the third highest concentration. This is a very common situation — ambient water vapor concentrations usually fall into the third position on the basis of decreasing component concentration for any air mass.

Flow Measurements vs. the Make-Up of the Gas Being Measured

One of the most common devices used to measure the flow rate of a gas is the rotameter, which is typically a transparent, circular cross section tube in which the inside diameter increases steadily as one travels upward in the tube. This rotameter tube will usually have graduated markings on its exterior surface, markings that permit a clear identification of any vertical position on that tube. In the center of the interior of this tube will be a "float" of some suitable material. This float is situated such that it can move freely upward and downward in its tube, in response to the passage of whatever gas is flowing in the rotameter. A rotameter then, as described here, is simply a variable area flow meter.

Rotameters are typically calibrated by their manufacturer. That is to say, the rotameter in question will be challenged under precisely known conditions of pressure and temperature with well-documented flows of some gas and the resulting position of the float noted. It is this tabulation, or the graphical representation of this tabulation, that constitutes the calibration of a rotameter. The most common conditions under which rotameter manufactures calibrate their rotameters are listed below — namely, STP conditions for either air or nitrogen as the calibrant gas.

> *Gas being Measured*   Air or Nitrogen
>
> *Barometric Pressure*   1 atmosphere or 760 mm Hg
>
> *Atmospheric Temperature*   0°C or 32°F

Whenever one must measure a gas *other than* air or nitrogen [for which the effective and actual molecular weights are, respectively: $MW_{air} =$ 28.964, and $MW_{nitrogen} = 28.013$ amu] a correction *must be made* — calibrations with either air or nitrogen are considered to be equivalent, since the effective and actual molecular weights [and densities] of these two fluids are virtually identical. The corrections that are required are based on the differences in the densities of the gases — i.e., the density of the "different fluid" or alternative gas that will be flowing in the rotameter vs. the density of the gas — either air or nitrogen — that was used to develop the initial rotameter calibration.

To understand the nature of this alternative gas correction, it is important to recognize that the flow rate that a typical user will want to achieve in his operations will be what the equivalent flow rate of the alternative gas would be if the operating conditions actually were to be *Standard Temperature and Pressure*. For the purposes of the formula on the following page, it will be assumed that whatever this alternative gas is, the conditions of the actual measurement under which the rotameter will be used will be Standard Temperature and Pressure. The goal of the correction formula, then, will be to identify the required rotameter *scale reading* that must be established to produce some specific *STP equivalent flow rate*.

The quantitative relationship is:

$$X = X_{\text{calibrated}}\sqrt{\frac{\rho}{\rho_{\text{standard}}}}$$

Where: X =   the required *scale reading* on a rotameter carrying a gas other than air or nitrogen, i.e., the *scale reading* that will be required in order to indicate that the flow rate of the alternative gas is numerically equivalent to some particular STP calibrated flow rate of either air or nitrogen;

$X_{\text{calibrated}}$ =   a specific *scale reading* obtained during the calibration of this rotameter — a reading that corresponded to a particular specific flow rate of the calibration gas, air or nitrogen, under conditions of STP;

ρ =   the density of the gas currently flowing through the rotameter, in some suitable set of units; and

$\rho_{\text{calibrated}}$ =   the density of the calibration gas — either air or nitrogen — in the same set of units as was the case for ρ.

To understand the specifics of this situation, let us consider an example.

Assume we are dealing with a rotameter that is to carry helium. Assume further that an investigator *must know* the specific rotameter *scale reading* that is required to produce an STP flow rate of 100.0 ml/min of helium in a rotameter that was calibrated using nitrogen under conditions of STP. Assume finally that the manufacturer's calibration of this rotameter showed that a flow rate of 100 ml/min of nitrogen occurred at a rotameter *scale reading* of 4.50 cm. To apply this formula we must know the values of the densities, at STP, for both the alternative gas, helium, and the calibration gas, nitrogen. In order, these densities are:

$$\rho_{\text{helium}} = 1.79 \times 10^{-4} \text{ grams/ml} = 0.179 \text{ grams/liter}$$

$$\rho_{\text{nitrogen}} = 1.25 \times 10^{-3} \text{ grams/ml} = 1.25 \text{ grams/liter}$$

Now applying the required formula we get:

$$X = X_{calibrated}\sqrt{\frac{\rho}{\rho_{standard}}} = (4.50)\sqrt{\frac{1.79 \times 10^{-4}}{1.25 \times 10^{-3}}}$$

$$X = (4.50)\sqrt{0.143} = (4.50)(0.378) = 1.70 \text{ cm}$$

What this means is that the rotameter float must be set at a *scale reading* of 1.70 cm if it is to deliver the required 100.0 ml/min STP equivalent flow rate of helium. Although everything stated in this example is completely correct and accurate, in actual practice, the situation is slightly different. To understand this, it is necessary to consider a typical rotameter.

Rotameters vary in overall length from less than 5.0 cm to more than 25 cm. They are always fabricated from some type of transparent material (i.e., glass, polycarbonate, etc.). This obviously is necessary in order to be able to observe the position of the rotameter float. Rotameters will always have their walls marked with labeled and uniformly spaced graduations — i.e., *scale readings* — in order to provide clear references as to the position of the float in any situation or circumstance. The major graduations on a typical rotameter will usually be measured in centimeters, with minor divisions at half or tenth centimeter distances. Obviously, the greeter the overall length of the rotameter, the lesser will be the distance between its minor divisions. Positioning a rotameter float at either one of its minor or its major divisions is a relatively easy process. Interpolating between minor *scale readings* to establish a float position at some non-graduated location is considerably more difficult.

In the example on the previous page, the desired *scale reading* was 1.70 cm. It is very likely that the rotameter being used would have 1/10 cm distances as its minor divisions. For such a rotameter setting the float at a *scale reading* of 1.70 cm would be a relatively simple matter. If the result of the calculation on the previous page had indicated that the *scale reading* had to be 1.72 cm, the situation would have proved to be very much more difficult. It is extremely challenging to center a rotameter float at a position that is *not* directly on one of the graduated marks on the rotameter body. Visual interpolation and setting of a rotameter float position to *scale readings* that are not directly on graduated major or minor division is process that is fraught with potential errors. Because of this, the actual process of setting the flow in a rotameter that is to carry a gas other than the one used to obtain the device's calibration is different than was described previously.

Specifically, the process usually involves carefully setting the float of the rotameter that is carrying an alternative gas on an easy-to-read *scale reading*, namely, one that is centered directly on one of the rotameter's graduated major and/or minor divisions. Once this has been accomplished, the STP flow rate of the alternative gas can be determined by a two step process. The first of these steps involves using the quantitative relationship from the previous page [slightly rearranged, as shown below] to determine what the equivalent calibrated *scale reading* would have been for the gas that was used to obtain the calibration of that specific rotameter. The rotameter's calibration data set is then used, in conjunction with the just calculated equivalent *scale reading* to determine the actual STP flow rate of alternative gas. In this second step, if interpolation is necessary [which is usually the case], the interpolation is mathematical rather than visual, and the potential for errors significantly reduced.

Again, let us consider an example.

Assume we are dealing with a rotameter that was calibrated using air but is to carry oxygen in the application being considered. Assume further that the manufacturer's air-based calibration data set for this rotameter — at STP — is as shown in this table.

Rotameter Calibration

Scale reading, cm	Flow Rate, ml/min
0.0	0
1.0	58
2.0	193
3.0	446
4.0	803
5.0	1,220
6.0	1,618
7.0	2,015
8.0	2,398
9.0	2,790
10.0	3,193

Assume, finally, that this oxygen carrying rotameter is set at a *scale reading* of 4.0 cm. The actual flow rate of oxygen is determined as follows — using, as stated above, a rearranged form of the relationship shown previously. This rearranged form is:

$$X_{calibrated} = X\sqrt{\frac{\rho_{standard}}{\rho}}$$

Again to apply this formula we must know the values of the densities, at STP, for both the alternative gas, oxygen, and the calibration gas, air. In order, these densities are shown next.

$$\rho_{oxygen} = 1.428 \times 10^{-3} \text{ grams/ml} = 1.428 \text{ grams/liter}$$

$$\rho_{air} = 1.292 \times 10^{-3} \text{ grams/ml} = 1.292 \text{ grams/liter}$$

Although these densities are fairly close together in value, there is enough difference between them to end up producing a different STP flow rate for the oxygen passing through the rotameter. Substituting into the listed formula, we get:

$$X_{calibrated} = X\sqrt{\frac{\rho_{standard}}{\rho}} = (4.0)\sqrt{\frac{1.292 \times 10^{-3}}{1.428 \times 10^{-3}}}$$

$$X_{calibrated} = (4.0)\sqrt{0.905} = (4.0)(0.951) = 3.81 \text{ cm}$$

Now, to determine the equivalent STP flow represented by an equivalent *scale reading* of 3.81 cm, we need only perform a simple linear interpolation between the calibrated flow rates for *scale readings* of 3.0 and 4.0 cm. From the calibration tabulation on the previous page, these flow rates are as follows:

at a scale reading of 3.0, the flow rate was 446 ml/min

at a scale reading of 4.0, the flow rate was 803 ml/min

The linear interpolation would be:

True Flow Rate $= 446 + [(0.81)(803 - 446)]$

True Flow Rate $= 446 + (0.81)(357) = 446 + 289.2 \approx 735$ ml/min

Clearly, the difference between the calibrated STP flow rate of air at a *scale reading* of 4.0 cm [803 ml/min], and the calculated true STP flow rate of oxygen at the same *scale reading* of 4.0 [735 ml/min] is significant — an actual difference of 68 ml/min or 9.25%.

Flow Measurements vs. Barometric Pressure and Temperature

The potential problem caused by making rotameter-based flow rate measurements at terrestrial altitudes other than Sea Level — i.e., at barometric pressures other than 760 mm Hg, at which pressure the rotameter was probably calibrated — is that *scale readings* at different altitudes will not correspond to the flow rates that were determined during the calibration. The same can be said for measurements that are made at temperatures other than 0°C. Basically the differences are directly analogous to the foregoing discussion on rotameters when they are carrying alternative gases.

The single factor that produces variations from the calibrated standard values in these situations will always be the density of the gas being transported through the rotameter vs. the density of the gas that was used to develop the calibration. In the discussion of the impact of having alternative gases flowing through the rotameter, the density factor was simply a function of the molecular weights of the materials involved, since the densities were always determined under the same ambient conditions, namely, STP.

If a rotameter is used to determine the flow rate of air or nitrogen (the same gases most commonly used to develop the rotameter's basic calibration) at an altitude other than Sea Level, and/or at a temperature other than 0°C, there may be a difference in the density of the air or the nitrogen flowing through the rotameter vs. the corresponding density at Sea Level. This density shift will have been caused either by the change in the prevailing barometric pressure, or by changes in the ambient temperature, or by both. This is a situation for which an alert analyst must always be ready to make adjustments.

The formula that permits these readings to be corrected has been produced on the following page. It is directly analogous to the previous relationship, namely, the one that involves alternative gases. It is based on accounting for the different gas density values; however, this time the determination of these differences is made under two different sets of ambient conditions rather than from two different gases. The mathematical relationship accomplishes these corrections directly from the ambient pressure and temperature, without requiring any additional density calculations, since we are dealing here with either air or nitrogen, the gases that were used to develop the STP calibration of the rotameter:

$$X = [0.5995][X_{standard}]\sqrt{\frac{P_{ambient}}{T_{ambient}}}$$

Where: X =   the required *scale reading* on a rotameter carrying air or nitrogen under temperature and pressure conditions different from those at STP;

$X_{standard}$ =   a specific *scale reading* obtained during the calibration of this rotameter — a reading that corresponded to a particular specific flow rate of the calibration gas, air or nitrogen, under conditions of STP;

$P_{ambient}$ =   the ambient barometric pressure, in mm Hg; and

$T_{ambient}$ =   the ambient temperature, in K.

Again, this relationship can probably be most easily understood by considering a specific example.

Suppose there is available a rotameter that has been calibrated using nitrogen at STP. For this rotameter, a calibrated float position of 23.0 mm corresponded to a nitrogen flow rate of 1,626 ml/minute. If this rotameter is to be used to identify the flow rate of air in Denver, CO, where: (1) the altitude is 5,280 feet above mean Sea Level, (2) the barometric pressure is 642.5 mm Hg, and (3) the ambient temperature is 32°C, what will be the required *scale reading* on this rotameter — the reading at which the flow rate of air at the Denver location would equal 1,626 ml/minute?

The solution:

We have been given the values of almost all of the parameters we will require. The only calculation we must make is to convert the ambient temperature, which has been given in °C, to its corresponding value in K. This is done simply by adding 273.16° to the ambient value of 32°C, thus:

$$T = 32° + 273.16° = 305.16 \text{ K}$$

Now, substituting into the formula, we can develop the required answer:

$$X = (0.5995)(23.0)\sqrt{\frac{642.5}{305.16}}$$

$$X = (13.79)\sqrt{2.106} = (13.79)(1.451) = 20.0 \text{ mm}$$

Therefore, the required *scale reading* on the rotameter located in Denver, CO — i.e., the reading that will correspond to a 1,626 ml/minute flow rate of air at that location — would be 20.0 mm.

Other Effects of Barometric Pressure and Temperature

Whenever one must make ambient concentration measurements of some chemical of interest at ambient temperatures and/or pressures other than those characteristic of STP, certain problems may develop. For the most part, these problems — like those related to the measurement of flow rates — will turn out to have been related to the density of the atmosphere in which the measurements are being made.

Most manufacturers calibrate their gas analyzers under conditions of Normal Temperature and Pressure (NTP), namely, 25°C and 760 mm Hg. Whenever an analyzer must be used under different ambient conditions, problems for certain analytical approaches may occur. To understand these potential problem areas, it will be instructive to consider the basic detection methodology of an infrared spectrophotometric gas analyzer. To that end, let us consider an example in which this type of analyzer is tasked to perform an analysis of the ambient concentration of acetone.

An IR spectrophotometer's operation is based on the relationship between the absorbance produced by some analyte of interest (as a component in the gas mixture being analyzed), to that analyte's concentration in the gas mixture. Simply, for any analyte, it is fairly easy to select an infrared wavelength where that analyte will absorb — i.e., for acetone, a wavelength of 1,208 cm^{-1} [8.28

μ], in the mid-infrared band, works very well. For an acetone analysis conducted under conditions of NTP, an infrared spectrophotometric analyzer, using the previously listed acetone absorbing wavelength, will be able to provide excellent results. The calibration on this type of analyzer, having been done at NTP, ensures that it will provide precise and accurate analyses of ambient acetone concentrations at NTP — i.e., it will respond perfectly to the population of acetone molecules that are characteristic of the acetone concentration as this population would exist in a matrix at NTP. Because this type of analyzer responds to a specific population of absorbing molecules, there will always be a potential problem for an analysis conducted under conditions other than those at which the analyzer was calibrated.

To understand this process, we must start by identifying the acetone molecule population, say at a concentration of 100 ppm(vol), in an air matrix at NTP. For such a situation, there will be a total of $2.46 \times 10^{21} = N_{acetone}$ acetone molecules/m³. Against its calibration at NTP, an infrared spectrophotometer will identify the absorbance produced by this number of molecules as corresponding to a concentration of 100 ppm(vol).

Assume next that we are dealing with an ambient acetone analysis in Denver, CO, on June 14, 1997, when the temperature was 88°F. Assume further that the acetone concentration in this Denver, CO, matrix was exactly 100 ppm(vol). Clearly, the ambient conditions in this situation were significantly different from those of NTP, specifically, the two sets of conditions, including data on the acetone molecule concentrations [both volume and mass-based], and the acetone molecule populations, are tabulated side-by-side below to emphasize the comparative differences:

Parameter	NTP Conditions	Denver, CO, Conditions 6/14/97
Altitude	Sea Level	5,280 feet above Sea Level
Ambient Temperature	25°C = 298.16 K	88°F = 31.11°C = 304.27 K
Barometric Pressure	760 mm Hg	642.5 mm Hg
True Acetone Conc.	100 ppm(vol)	100 ppm(vol)
Indicated Acetone Conc.	100 ppm(vol)	82.9 ppm(vol)
True Acetone Conc.	237.5 mg/m³	196.9 mg/m³
Acetone Molecule Count	2.46×10^{21} molecules/m³	2.04×10^{21} molecules/m³

For this matrix, with its reduced number of acetone molecules per unit volume, but with the same volumetric acetone concentration, an infrared spectrophotometer, calibrated under conditions of NTP, would produce a measured acetone concentration of 82.9 ppm(vol). Clearly there is a problem.

The basic nature of the problem in this situation again relates to the density of the gas in the matrix where the measurements are being made. The formula that will permit the correction of this problem is

$$C_{true} = 2.549 C_{indicated} \left[\frac{T_{ambient}}{P_{ambient}} \right]$$

Where: C_{true} =   the true concentration of the analyte of interest under the prevailing, non-NTP conditions of ambient temperature and pressure;

$C_{indicated}$ =   the concentration indicated by an analyzer for which the principle of detection is dependent upon the analyte's actual molecular population per unit volume in the matrix being monitored;

$P_{ambient}$ =   the ambient barometric pressure, in mm Hg; and

$T_{ambient}$ =   the ambient temperature, in K.

It remains now only to list some of the analytical methods for which this corrective formula will be applicable. Seven of the common ones are listed in Table 11.

Table 11.   Analytical methods using the corrective formula

No.	Monitoring Technology	Correction Formula Applicable
1	IR Spectrophotometry	Yes
2	NDIR	Yes
3	Photo Ionization Detectors	Yes
4	Flame Ionization Detectors	Yes
5	Electrochemical Detectors	Yes
6	Solid State Detectors	Yes
7	Catalytic Bead Detectors	Yes

Appendix D
Periodic Table of Elements

The new IUPAC format numbers the groups from 1 to 18. The previous IUPAC numbering system and the system used by Chemical Abstracts Service (CAS) are also shown. For radioactive elements that do not occur in nature, the mass number of the most stable isotope is given in parentheses.

References
1 G J Leigh, Editor. *Nomenclature of Inorganic Chemistry*. Blackwell Scientific Publications. Oxford. 1990
2 *Chemical and Engineering News*. 63(5), 27. 1985
3 Atomic Weights of the Elements. 1995. *Pure & Appl. Chem.* 68, 2339, 1996

Appendix E
Standard Atomic Weights (1995)

At. no.	Name	Symbol	Atomic weight	At. no.	Name	Symbol	Atomic weight
1	Hydrogen	H	1.00794(7)	57	Lanthanum	La	138.9055(2)
2	Helium	He	4.002602(2)	58	Cerium	Ce	140.116(1)
3	Lithium	Li	6.941(2)	59	Praseodymium	Pr	140.90765(2)
4	Beryllium	Be	9.012182(3)	60	Neodymium	Nd	144 24(3)
5	Boron	B	10.811(7)	61	Promethium	Pm	[145]
6	Carbon	C	12.0107(8)	62	Samarium	Sm	150.36(3)
7	Nitrogen	N	14.00674(7)	63	Europium	Eu	151.964(1)
8	Oxygen	O	15.9994(3)	64	Gadolinium	Gd	157 25(3)
9	Fluorine	F	18.9984032(5)	65	Terbium	Tb	158.92534(2)
10	Neon	Ne	20.1797(6)	66	Dysprosium	Dy	162.50(3)
11	Sodium	Na	22.989770(2)	67	Holmium	Ho	164.93032(2)
12	Magnesium	Mg	24.3050(6)	68	Erbium	Er	167.26(3)
13	Aluminum	Al	26.981538(2)	69	Thulium	Tm	168.93421(2)
14	Silicon	Si	28.0855(3)	70	Ytterbium	Yb	173.04(3)
15	Phosphorus	P	30.973761(2)	71	Lutetium	Lu	174.967(1)
16	Sulfur	S	32.066(6)	72	Hafnium	Hf	178.49(2)
17	Chlorine	Cl	35.4527(9)	73	Tantalum	Ta	180.9479(1)
18	Argon	Ar	39.948(1)	74	Tungsten	W	183.84(1)
19	Potassium	K	39.0983(1)	75	Rhenium	Re	186.207(1)
20	Calcium	Ca	40.078(4)	76	Osmium	Os	190.23(3)
21	Scandium	Sc	44.955910(8)	77	Iridium	Ir	192.217(3)
22	Titanium	Ti	47.867(1)	78	Platinum	Pt	195.078(2)
23	Vanadium	V	50.9415(1)	79	Gold	Au	196.96655(2)
24	Chromium	Cr	51.9961(6)	80	Mercury	Hg	200.59(2)
25	Manganese	Mn	54.938049(9)	81	Thallium	Tl	204 3833(2)
26	Iron	Fe	55 845(2)	82	Lead	Pb	207.2(1)
27	Cobalt	Co	58.933200(9)	83	Bismuth	Bi	208 98038(2)
28	Nickel	Ni	58.6934(2)	84	Polonium	Po	[209]
29	Copper	Cu	63 546(3)	85	Astatine	At	[210]
30	Zinc	Zn	65 39(2)	86	Radon	Rn	[222]
31	Gallium	Ga	69.723(1)	87	Francium	Fr	[223]
32	Germanium	Ge	72.61(2)	88	Radium	Ra	[226]
33	Arsenic	As	74.92160(2)	89	Actinium	Ac	[227]
34	Selenium	Se	78.96(3)	90	Thorium	Th	232.0381(1)
35	Bromine	Br	79.904(1)	91	Protactinium	Pa	231 03588(2)
36	Krypton	Kr	83.80(1)	92	Uranium	U	238.0289(1)
37	Rubidium	Rb	85.4678(3)	93	Neptunium	Np	[237]
38	Strontium	Sr	87.62(1)	94	Plutonium	Pu	[244]
39	Yttrium	Y	88.90585(2)	95	Americium	Am	[243]
40	Zirconium	Zr	91.224(2)	96	Curium	Cm	[247]
41	Niobium	Nb	92.90638(2)	97	Berkelium	Bk	[247]
42	Molybdenum	Mo	95.94(1)	98	Californium	Cf	[251]
43	Technetium	Tc	[98]	99	Einsteinium	Es	[252]
44	Ruthenium	Ru	101.07(2)	100	Fermium	Fm	[257]
45	Rhodium	Rh	102.90550(2)	101	Mendelevium	Md	[258]
46	Palladium	Pd	106.42(1)	102	Nobelium	No	[259]
47	Silver	Ag	107.8682(2)	103	Lawrencium	Lr	[262]
48	Cadmium	Cd	112.411(8)	104	Rutherfordium	Rf	[261]
49	Indium	In	114.818(3)	105	Dubnium	Db	[262]
50	Tin	Sn	118.710(7)	106	Seaborgium	Sg	[263]
51	Antimony	Sb	121.760(1)	107	Bohrium	Bh	[262]
52	Tellurium	Te	127.60(3)	108	Hassium	Hs	[265]
53	Iodine	I	126.90447(3)	109	Meitnerium	Mt	[266]
54	Xenon	Xe	131.29(2)				
55	Cesium	Cs	132.90545(2)				
56	Barium	Ba	137.327(7)				

Numbers in brackets are the mass numbers of the longest-lived isotope of elements for which a standard atomic weight cannot be defined

Index

Printed and bound by CPI Group (UK) Ltd, Croydon, CR0 4YY

22/10/2024

01777624-0008